Anne Frank's Tree

ANNE FRANK'S TREE

NATURE'S CONFRONTATION WITH TECHNOLOGY, DOMINATION, AND THE HOLOCAUST

Eric Katz

New Jersey Institute of Technology

The White Horse Press

Copyright © 2015

The White Horse Press, 10 High Street, Knapwell, Cambridge, CB23 4NR, UK

Set in 11 point Adobe Caslon Pro
Printed by Lightning Source

All rights reserved. Except for the quotation of short passages for the purpose of criticism or review, no part of this book may be reprinted or reproduced or utilised in any form or by any electronic, mechanical or other means, including photocopying or recording, or in any information storage or retrieval system.

British Library Cataloguing in Publication Data
A catalogue record for this book is available from the British Library
ISBN   978-1-874267-85-0 (HB)
ISBN   978-1-874267-91-1 (PB)

~ Contents ~

Acknowledgements vi

Preface
　The Tree 1

Chapter One
　The Warsaw Cemetery and the Liberation of Nature 14

Chapter Two
　Thoughts on the Holocaust in the Spanish Synagogue of Venice:
　Human History, Technology, and Domination 38

Chapter Three
　Ecological Restoration and Domination:
　The Need for an Independent Nature 71

Chapter Four
　Independent Nature Denied 112

Chapter Five
　The Dark Side of Authenticity:
　Nativism, Restoration, and Genocide 138

Chapter Six
　Ethical Coda:
　The Nazi Engineers and Technological Ethics in Hell 164

Epilogue
　Fire Island, July 2012 180

References 183

Index 191

Acknowledgements

I have been thinking and writing about the subjects in this book for twenty-five years. I have spoken to many people, attended many conferences, and corresponded with scores of colleagues and students about my ideas. It is impossible to remember them all, much less thank them all by name. Yet I do wish to thank you all—anyone who has ever discussed with me ideas about ecological restoration, domination, the value of nature, the philosophy of technology, or the Holocaust.

A few people stand out in my memory as being particularly important:

Roger Gottlieb, who in 1990 organized a joint session of the Radical Philosophy Association, The Society for the Philosophical Study of Genocide and the Holocaust, and the International Society for Environmental Ethics (of which I was Vice-President) on the subject "Genocide and Eco-cide." Holmes Rolston thought that I should represent the ISEE, and so I wrote my first paper that remotely anticipated the themes of this book and began a decades-long study of the issues.

Bron Taylor, who invited me to give a keynote lecture at a conference in Amsterdam for the International Society for the Study of Religion, Nature, and Culture in July 2009; that lecture became the paper, "Anne Frank's Tree: Thoughts on Domination and the Paradox of Progress" which served as the basis of this book.

Steven Vogel, Eric Higgs, Ned Hettinger, and Yeuk-Sze Lo, for being steadfast critics and interlocutors of my position on ecological restoration. Also Piers H. G. Stephens for important comments and suggestions regarding my research into Nazi environmental policy and the value of forests. Eugene Hargrove of the journal *Environmental Ethics* has always been a major supporter of my ideas.

I would like to thank the administration and my colleagues in the Department of Humanities at the New Jersey Institute of Technology, especially Provost Fadi Deek. Much of the research and writing of this book was done during two sabbatical leaves from the university, for which I am extremely grateful. A special thanks should go to Bob Lynch who graciously read the entire first draft of this book, offering a multitude of suggestions (many of which I ignored).

Andrew Johnson and Sarah Johnson at White Horse Press, for having the courage to publish this book that defies all attempts at disciplinary categorization.

And finally my long time colleagues and friends, Avner de-Shalit, Andrew Light, and Andrew Brennan—you guys know why.

• • •

Grateful acknowledgement is made to the following publications for permission to reprint sections of these works in this book.

Parts of the Preface are derived from "Anne Frank's Tree: Thoughts on Domination and the Paradox of Progress," *Ethics, Place and Environment* (2010) **13**: 283–293.

Chapter One is derived from "Nature's Presence: Reflections on Healing and Domination," in Andrew Light and Jonathan M. Smith (eds), *Philosophy and Geography I: Space, Place, and Environmental Ethics* (Lanham, MD: Rowman & Littlefield, 1996), pp. 49–61, and reprinted in Eric Katz, *Nature as Subject: Human Obligation and Natural Community* (Lanham, MD: Rowman & Littlefield, 1997), pp. 189–201, and "The Liberation of Humanity and Nature," *Environmental Values* (2002) **11**: 397–405.

Chapter Two is derived from "The Authenticity of Place in Culture and Nature: Thoughts on the Holocaust in the Spanish Synagogue of Venice," *Philosophy & Geography* (2002) **5**: 195–204 and "On the Neutrality of Technology: the Holocaust Death Camps as a Counterexample," *Journal of Genocide Research* (2005) **7**: 409–421.

Chapter Three is derived from "Further Adventures in the Case Against Restoration," *Environmental Ethics* (2012) **34**: 67–97.

Parts of Chapter Four are derived from "Anne Frank's Tree: Thoughts on Domination and the Paradox of Progress," *Ethics, Place and Environment* (2010) **13**: 283–293.

Chapter Five is derived from "The Nazi Comparison in the Debate Over Restoration: Nativism and Domination," *Environmental Values* (2014) **23**: 377–398, and from "Anne Frank's Tree: Thoughts on Domination and the Paradox of Progress," *Ethics, Place and Environment* (2010) **13**: 283–293.

Chapter Six is derived from "The Nazi Engineers: Reflections on Technological Ethics in Hell," *Science and Engineering Ethics* (2011) **17**: 571–582.

This book is dedicated to

EMMA, ANI, JONAH

and

SUSAN

～ Preface ～

THE TREE

Our chestnut tree is in leaf, and here and there you can already see a few small blossoms.

From The Diary of Anne Frank

I.

Consider a tree that once grew in Amsterdam. There was a horse chestnut that grew behind the Anne Frank house, before it succumbed to old age, disease, and a severe windstorm in August 2010. It was the tree that Anne wrote about in her diary. As reported in *The New York Times* over a year before, the city of Amsterdam had abandoned plans to remove the tree—sickened by fungi and moths—in 2007, and instead had stabilized it with structural supports so that it could stand, it was hoped, for another fifteen years. More interesting, however, a plan was devised to keep the tree alive in perpetuity, by taking ten saplings that came from the original tree and planting them in various sites around the world—including one at the White House and one at the National September 11 Memorial in New York. The trees, according to the Anne Frank Center USA, will be a symbol of the growth of tolerance.

In this book, I use Anne Frank's relationship with this horse chestnut tree as a new starting point for an examination of post-enlightenment ideas about nature, domination, autonomy, technology, and human evil. Trees, and the forests and landscapes in which they appear, have always represented for me the natural world, the forces and processes that constitute nature. The existence of trees and their connection to human beings—to human projects and human institutions—can tell us what the natural world means for humanity, can tell us what value lies within the natural world. If we view forests as "lumber plantations" that provide humanity with wood for a variety of purposes, then nature is a mere resource, an instrument for the furtherance of human ends. If we view forests as parks or national monuments, then nature is a realm for human recreation. If we view forests as a pristine wilderness, then perhaps nature is a source of wonder and awe, a repository of the sacred and the ineffable, a world that exists apart from the mundane realities of human life. Perhaps a nature separate from human projects is an expression of autonomy, of freedom and self-realization.

Trees can also be tied to the idea of domination. I have long argued that the process of ecological restoration, in which a kind of environmental engineering attempts the re-creation of previously degraded or destroyed natural environments, is an example of the human project to assert our technological mastery over the autonomous processes of the natural world.[1] The management of forests for human purposes—whether these purposes are economic, aesthetic, environmental, or spiritual—is a type of domination. Whatever the situation, humans are imposing their wills onto the natural world to effect a change or produce a result that will achieve human goals. Thus, even the obviously eco-friendly or green activity of planting trees as part of the project of sustainable development can be conceived as the human attempt to dominate and control the natural world.

Nature, however, is perfectly capable of dominating humanity. The most cogent example, perhaps, is the weather, especially the extreme storms and abnormal temperatures that have been visited on many parts of the world in recent years: not only the devastation caused in New York and New Jersey by Hurricane Sandy but also the violent tornado-producing rainstorms in the American mid-west, the cruelly freezing winter of 2011–2012 in Eastern Europe, and the East African drought of 2011–2012, the worst in sixty years. Even before Sandy, I have witnessed damaging coastal storms that attacked Fire Island, close to my home, and I have considered these acts of nature to be akin to human aggression and imperialism.[2] But the power of nature can also heal the wounds produced by human activity. Consider the trees that grew in the Jewish cemetery of Warsaw, essentially unattended for decades after the Second World War—trees that I will discuss in the next chapter. I have seen how these trees grew into a forest over the gravestones and unmarked mass graves.[3] The natural growth of these trees in that place can begin to teach us about the essence of human evil and its relationship to the healing power of nature.

One cannot observe the trees of the Jewish cemetery of Warsaw or the horse chestnut behind the Anne Frank house without, of course, contemplating the role of the Nazis in the creation of human evil. The

The Tree

actions and policies of the Third Reich constitute a major part of this book, for the Nazi regime is perhaps the finest example of the project of domination. Yet this is not a book, not an argument, about the death camps and genocide. The starting point of this book is the existence of trees—and one tree in particular—and so it is nature, and its domination by human technology, that is the primary focus. So in one sense this is a book of environmental philosophy, a book about environmental ethics and the proper human relationship to the natural world. But we cannot avoid the fact that the Anne Frank tree is only important to us today because of its past relationship to the Third Reich. When we contemplate the meaning of this tree in its historical context, what is compelling about the Nazi regime is the interconnection and inescapable relationship between the domination of nature and the domination of humanity, for the environmental policies of the Nazis were eventually co-extensive with their policies of genocide. The Third Reich attempted to re-make the world—or at least Europe—into its ideal of a pure Aryan homeland, a natural and cultural landscape purified of all non-German elements. This project involved not only the murder of millions of non-Germans but also the control and manipulation of the natural environment. The Nazi regime presents us with a seamless connection between the domination of nature and the domination of humanity. Thus this is a book that moves beyond the traditional realm of environmental philosophy and into the realm of human history, the history and meaning of the human domination of nature and humanity.

    At the heart of the project of domination is the power of human technology. It is technology in various forms—agriculture, biological science, environmental engineering, to cite just a few examples—that is responsible for the re-making of the natural world into a cultural landscape amenable to humanity. In almost all instances, the intentions behind the use of the technology are beneficent, meant for the production of good. It was human technology, after all, that was employed in the vain attempt to preserve the Anne Frank tree. The powers of nature, however, were strong enough, or uncontrollable enough, to subvert the plans of humanity in that case. Despite the implementation of structural supports, the tree did not survive a fierce windstorm. But technology is also being used for the sapling project, and it remains to be seen how successful this use of agricultural or gardening technology will be. Even

Preface

if the sapling project is successful on a physical level, in that various trees derived from the original Anne Frank tree grow and flourish in designated sites around the world, the use of technology on the original tree will affect the meaning that we give to the descendants that survive. These new saplings will be artifactual products that we have created to replace, in some sense, the original tree that inspired Anne Frank. They will have a different meaning than the original tree, and this meaning will be imposed on the trees by a contemporary human population that has a vastly different agenda, a different set of goals, than Anne Frank, a teenaged Jewish girl hiding from the Nazi regime.

The use of technology changes the meaning of natural processes. Indeed, one of the principal themes of this book is that technology is a form of human domination. Technology is a physical manifestation of human intentionality imposed upon the world. When this technological domination is imposed upon the natural environment—and this is a second principal theme—humanity creates an artifactual world that merely resembles nature. A nature created by human technology and science is an illusion, a mere cheat that hides and covers up the harmful consequences of human activity. Human activity, of course, may be the result of good or evil intentions, yet a third principal theme of this book is the presence of human evil, here represented mostly by the actions of the Third Reich. The problem with the technological domination of nature is that the ends pursued by humanity are subject to the prevailing ideologies of specific political regimes, and there may be no way to insure that these ends are morally worthwhile.

Thinking about the Anne Frank tree and the sapling project offers an opportunity to examine the interconnections between nature, technology, the ideas of domination and autonomy, and the human evil of the Holocaust. Should the saplings be a symbol of tolerance, as has been proposed? Or can these trees be a symbol of resistance? For me, Anne Frank used her horse chestnut tree to create a force to confront the powers of human evil that surrounded her. In nature—in this tree—she found a touchstone to resist the continual human project of the domination of nature and humanity.

The Tree

II.

Begin with what Anne Frank wrote in her diary about this tree. For all of its current renown, the tree plays a small part in the diary; it is not even mentioned until the entry of February 23, 1944, the nineteenth month of the family's hiding; a little over five months later the secret annex will be discovered and everyone will be sent to the death camps. In February of 1944, fourteen-year-old Anne is deeply infatuated with Peter van Pels (the teenage boy of the other family in hiding in the annex) and she spends as much time as possible in his attic room. She writes about this day:

> The two of us looked out at the blue sky, the bare chestnut glistening with dew, the seagulls and other birds glinting with silver as they swooped through the air, and we were so moved and entranced that we couldn't speak ... "As long as this exists," I thought, "this sunshine and this cloudless sky, and as long as I can enjoy it, how can I be sad?" The best remedy for those who are frightened, lonely, or unhappy is to go outside, somewhere they can be alone, alone with the sky, nature and God. For then and only then can you feel that everything is as it should be and that God wants people to be happy amid nature's beauty and simplicity ... I firmly believe that nature can bring comfort to all who suffer.[4]

The beauty of the tree is immediately connected to the overall processes of nature, and even God's plan to create happiness and relieve the suffering of human beings through the processes of nature. Anne repeats this idea several times: "My advice is: 'Go outside, to the country, enjoy the sun and all nature has to offer. Go outside and try to recapture the happiness within yourself.'"[5] One April day, Anne writes of her first kiss, and now the praise of nature intensifies:

> After our mild winter we've been having a beautiful spring. April is glorious, not too hot and not too cold, with occasional light showers. Our chestnut tree is in leaf, and here and there you can already see a few small blossoms ... What could be nicer than sitting before an open window, enjoying nature, listening to the birds sing, feeling the sun on your cheeks, and holding a darling boy in your arms?[6]

And indeed—is it because of that darling boy?—this year, the chestnut "is even more beautiful than last year."[7]

Nature thus becomes for Anne a countervailing force to oppose the horror of her life in hiding and the oppressive human world that has created the conditions that make this hiding necessary. The chestnut tree is merely one symbol of this nature, for she also mentions the snow that has fallen,[8] the moon,[9] as well as the birds, the sun, and the sky. Nature serves as a sign that good still exists in the world. Indeed, in the paragraph that follows the best known lines of the diary—"I still believe, in spite of everything, that people are truly good at heart"—she returns to Nature as the source of hope amid a world bent on destruction and death: "I hear the approaching thunder that, one day, will destroy us, too, I feel the suffering of millions. And yet, when I look up at the sky, I somehow feel that everything will change for the better, that this cruelty too will end, that peace and tranquility will return once more."[10]

This hope in Nature is necessary because of the destructive evil of the human race. When Anne considers the reasons and causes for the war that is being waged around her, she attributes the blame to the nature of humanity itself:

> There's a destructive urge in people, the urge to rage, murder and kill. And until all of humanity, without exception, undergoes a metamorphosis, wars will continue to be waged, and everything that has been carefully built up, cultivated and grown, will be cut down and destroyed, only to start all over again![11]

Humans hate, destroy, and kill; Nature, as we have seen from other diary entries, inspires happiness, a love of beauty, and peace. Anne wonders if it is only because she has been cut off from the world that she has developed a newer and deeper appreciation of Nature:

> Is it because I haven't been outdoors for so long that I've become so smitten with nature? I remember a time when a magnificent blue sky, chirping birds, moonlight and budding blossoms wouldn't have captivated me. Things have changed since I came here ... It's not just my imagination—looking at the sky, the clouds, the moon and the stars really does make me feel calm and hopeful ... Nature makes me feel humble and ready to face every blow with courage! ... Nature is the one thing for which there is no substitute![12]

Nature serves as a source of hope for peace and happiness against the destructive power of human evil. And it is Nature alone that cannot be replaced—"there is no substitute." Anne Frank herself tells us how we should understand the meaning of the chestnut tree, and its ten new saplings being planted around the globe: the tree represents the unique and irreplaceable power of Nature to confront the human evils of oppression and domination. The tree must be seen as a symbol of resistance.

The Tree

III.

In reading these passages of the diary, we can see that Anne Frank's chestnut tree is connected to the idea of domination, and, ultimately, to complex ideas about the essence of human progress and liberation. I consider domination to be the fundamental evil that underlies human relationships—relationships both to the natural world and to other human beings, human communities, and human institutions. In my previous work, I have tried to draw parallels between the evil treatment of human beings and the evil treatment of nature—thus I think of Nature as analogous to a human subject, susceptible on one hand to the evils of oppression and domination, and on the other hand to the goods of freedom and autonomy.[13] This analogy—nature as subject—is a radical idea, quite different from the traditional view of nature as a merely physical collection of forces, material, and processes. It permits an expansion of our evaluation of nature. It allows us to see nature as an entity that develops or unfolds, so to speak, according to an internal logic. By claiming that nature is analogous to a human subject, I am not claiming that nature possesses a consciousness or intentionality: nature does not know what it does, nor does it plan its actions as a human rational agent would. But nature does act with a kind of autonomous freedom, unless it is managed or controlled by human technology. We can witness this autonomous unfolding of natural processes, and value it, without any commitment to a theory of the teleology of nature. The value of an autonomous nature does not rely on the achievement of some pre-ordained end of natural processes, but in the mere fact that nature develops on its own, without the interference of human projects, plans, and control. An autonomous nature thus stands in opposition to domination, first in opposition to the attempted management of the natural world, but second, and more importantly, in opposition to the oppression of humanity—as I will argue in this book.

Anne's tree is a symbol of this autonomous nature, opposed to domination, grounded in an historical context. It thus becomes a symbol of resistance, but again, a resistance tied to a feature of a specific historical epoch: the oppression of the Third Reich. For Anne, the tree represents a nature that gives her courage "to face every blow" of the human evil that surrounds her. And it is clear that the reason for this meaning is the autonomous development and freedom of the natural processes; the birds,

Preface

the blossoms, the clouds, and sky all continue to exist and to function regardless of the evil human activities that threaten the destruction of civilization. More important, especially for the argument of this book, is the idea that the tree is a representative of a natural world that cannot be replaced—again, "nature is the one thing for which there is no substitute." Nature alone can give us the strength to resist the worst offenses of humanity. Although Anne believes that "people are truly good at heart," she also believes "there's a destructive urge in [them]" that must be changed completely before peace can flourish on this planet. So it is nature, not people, that is the source of hope for Anne. The irreplaceable natural world—this horse chestnut tree—must continue to exist to counteract the evil that men do.

Nature, domination and autonomy, and the presence of human evil: these again are the topics of this book. In the following chapters, I will build on these introductory ideas about the Anne Frank tree to explore broader issues in our ethical relationships to both nature and humanity. The human use of technology lies at the heart of these relationships, for the control and domination of both nature and humanity are advanced by the inappropriate (and sometimes, vicious) employment of technological processes. Nature, domination, and technology are inextricably linked, and the fullest expression of an evil fusing of these ideas can be found in the Nazi relationship to nature and humanity. It is the task of the Anne Frank tree to shield us from this evil, to present an alternative, to help us resist the forces of domination. This, then, is a book of environmental philosophy or environmental ethics, but it is unlike any other book in the field, because it ties together the idea of the domination of nature with the history of Nazism.

In Chapter One, I will begin with a visit to the Jewish cemetery in Warsaw and the Majdanek death camp near Lublin. In the cemetery the vegetation has grown unattended for decades after the Second World War, covering the graves of murdered Jews with a lush forest of undergrowth and trees. In the death camp the grass has covered the fields of mud and the remains of the ashes. Here we can observe another intersection of nature and the forces of domination, yet the result seems to be beneficial, for the processes of nature have obscured the most obvious signs of human evil. Is nature powerful enough to heal the violence perpetrated by humanity? I believe that this question represents a point of view that may

The Tree

reinforce the very processes of domination we need to resist. My argument begins with the personal experience of these Holocaust sites and their connection to nature as a means for opening up the meaning of human domination; I argue that in regard to nature, human domination is an expression of anthropocentrism, the privileging of human interests and benefits. An anthropocentric worldview is pervasive in environmental thought and practice, as we can see in the policy of sustainable development or the process of ecological restoration. In this chapter, I introduce a critical evaluation of restoration, based on two decades of my previous work, as a central theme of this book. The restoration of nature by human science and technology results in the creation of a world comprised of artifacts. But the healing of sites of human evil by natural forces is the mirror image of the process of ecological restoration: it cannot restore a prior reality that has been irrevocably altered. Understanding nature as a healer of human evil is to understand nature as an anthropocentric agent of human interests. I argue, conversely, that we should view the power of nature to cover up the wounds caused by humanity as a demonstration of nature's independence and autonomy. Nature liberated from human management and control develops in a way that is divorced from human interests and concerns. Nature does not care that the trees are covering the mass graves of murdered Polish Jews. For me, this independence of nature is the source of its value, the reason why we humans should wish to preserve and protect it. The autonomy of nature, its liberation from human domination, should be the central goal of human activity regarding the natural world, just as the liberation of humans, their freedom from oppression, should be our central goal regarding humanity.

The connection of human history to the processes of domination—in both the spheres of humanity and of nature—is the major subject of Chapter Two. Here we begin with a visit to the Spanish synagogue of Venice, Italy. In this synagogue are the descendants of Jewish families that survived the Holocaust of European Jewry perpetrated by the Nazi regime of the Third Reich. Sitting in the Spanish synagogue of Venice, I am aware of the overwhelming importance of human history in the understanding of destruction and domination. Whether we consider the degradation and abuse of the natural environment, or the oppression and genocide of human peoples and cultures, we cannot avoid the historical context as the basis of analysis and meaning. In this chapter we will also consider the operations of the Nazi death camps and the traditional argument about the ethical neutrality of technology. An understanding of the technological history of the death camps such as Auschwitz-Birkenau and Majdanek demonstrates the fallacy of this traditional argument. In the camps we see a technology that embodies a specific set of moral and political values. We see the power of technology to re-make the world according to a specific human ideology.

The role of technology in the literal creation of a new world leads us back to the process of ecological restoration, the subject of Chapter Three. Here I review and expand on arguments concerning the moral value and meaning of the restoration process that have been debated in the environmental philosophy literature of the past twenty years. In my view, restoration ecology is a continuation of the paradigm of human scientific and technological mastery over natural processes. The underlying technological assumption is that humans can control natural processes to better effect than nature can. This will change the essential character of environmental policy, moving it away from preservation and protection, and replacing it with manipulation and control. The debate over restoration ecology raises significant issues about the meaning of nature and the place of humanity in natural processes. Here I argue for the necessity of a human/nature dualism as a prerequisite for understanding the meaning of environmental policy. Nature must be conceived as distinct from human projects and institutions, for as Anne Frank so pithily expressed it in her diaries, "nature is the one thing for which there is no substitute." The conclusion is that the process of ecological restoration—which attempts to replace or substitute an alternative human creation for nature—is a mistaken and potentially destructive goal of environmental policy.

Before we see how destructive a policy of restoration can be (in Chapter Five), Chapter Four considers a possible objection to the entire framework that structures my argument, namely, that the idea of an independent nature existing outside of human institutions is a fallacy. As an introduction to this objection, I use Simon Schama's extensive discussion of the history of forest landscapes in Europe and America.[14] Schama claims that landscapes are cultural products, or projections, heavily influenced by metaphor. The metaphors we use to understand nature become the reality, more real than the actual entities themselves. This is

The Tree

a form of social constructivism, the theory that human ideas shape, and even create, the world in which we live. It is true that different national and regional narratives inform and determine differing relationships to local and regional natural landscapes, but for Schama (and other social constructivists), the human relationship to the landscape defines the landscape, defines nature in its interaction with humanity, so that the reality of an independent nature is yet another myth or metaphor. An independent nature has no reality outside of the conscious meanings of humanity. For me, the social constructivist view espouses another form of the human domination of nature: call this "epistemological domination." Epistemological domination is perhaps the necessary condition for the physical domination of nature: it assumes that not only our science and technology can master Nature, but our ideas can master it as well. This means that we can alter and manipulate reality while at the same time we can believe confidently that we are respecting the integrity of natural processes. A consideration of Anne Frank's tree and the independent nature it represents can help us answer the social constructivist objection posed by Schama: we will see that although society does, in part, determine the meaning of landscape, environment, and nature, there exists a physical reality that transcends humanity, human thought, and human institutions.

In Chapter Five we examine the convergence of these arguments about the power of nature, restoration and human technology, history and metaphor, and the oppressive forces of human evil through domination by an examination of Nazi environmental policy and its similarities to specific claims of some advocates of ecological restoration. The evil that confronted Anne—the Nazi Third Reich and its dreams of domination—is a story connected to trees and nature. The Nazi plans for expansion and domination were intertwined with mythic narratives about the German connection to the land and nature. "Blood and soil"—Richard Darré's catchphrase for the racial purity and superiority of authentic Germans—had its roots in the German forests and peasant agricultural practices. But the concept of blood and soil—the national and racial connection to the natural landscape—is a much more complex story than it first appears. Throughout the Nazi period in Germany, there is a strong predilection to create a new kind of *volkisch* forest policy opposed to the traditional German scientific forestry of the nineteenth century. This

Preface

new forestry, called the *Dauerwald* concept ("eternal forest"), had many distinctive elements, but one leading characteristic was an emphasis on pure or native species, and the elimination of exotics that were deemed alien to the German landscape. Contemporary restoration ecology often echoes these nativist claims, and arguments against exotic species are tied to arguments about the dangers of the globalization of natural and regional ecosystems and the loss of biodiversity. The connection to eliminationist sentiments is, of course, disturbing, but I argue in this chapter that the fundamental evil is not the echoes of Nazi-like eliminationist and genocidal rhetoric but rather the existence of domination and oppression. Restoration ecology is not evil because it attempts to eliminate native species; it is evil because it continues the paradigm of human technological mastery over autonomous natural processes. And once again, it is a belief in an independent and autonomous natural world, standing outside the technologically managed human landscape, which can be used to resist the oppressive power of human domination. The symbol of this autonomous natural world, the symbol of this resistance, is Anne Frank's tree.

Chapter Six, the final chapter of the book, is a concluding coda that takes us away from the natural environment and into the heart of the ethical decisions made by the creators of Nazi technology. We have been dealing with human activity in the natural environment and how the human attempt to control nature is connected to the control of other human beings. In the death camps, obviously, the oppression and domination of human beings reaches its apex in the systematic technologically based genocide of those people deemed subhuman by the ideology of Nazism. We could examine the direct perpetrators of this genocide, the SS officers and guards in the camps, but I believe it is more interesting and significant to examine the ethical decisions and actions of those who ought to have had a higher social responsibility: the scientific and technological professionals. How did the engineers that designed the death camps of the Third Reich evaluate the moral dimensions of their genocidal creations? Although this topic appears, on the surface, to be far removed from the discussion of Nazi environmental policy and the human domination of nature, it is a necessary conclusion to the overall argument. It examines a detailed case study in the process of ethical deliberation. To design and create a technology that is appropriate to

The Tree

the continued existence of the natural world, an engineer must know that the technology that is created is a good technology. But how does the engineer know this? How does an engineer know that the values he embodies through his technological creations are good values that will lead to a better world? The Nazi engineers that designed the gas chambers and crematoria believed that they were creating a better world. We need to understand why they endorsed policies of oppression and genocide. Only then can we begin to resist the ideology of domination, and use this resistance to create a better world for both nature and humanity.

Notes to Preface

1. See Katz 1991; 1992a; 1992b; 1993; 1995; 1996a; 1996b; 1997; 2000; 2002; 2009; 2010; 2011; 2012.
2. Katz 1995; 1997.
3. Katz 1996a; 1997.
4. Frank 1991, 196–97: 23 February 1944. All references to the Anne Frank diary are to this volume, edited by Otto H. Frank and Mirjam Pressler and translated by Susan Massotty.
5. Frank 1991, 211: 7 March 1944.
6. Frank 1991, 270: 18–19 April 1944.
7. Frank 1991, 297: 13 May 1944.
8. Frank 1991, 206: 4 March 1944.
9. Frank 1991, 227: 20 March 1944.
10. Frank 1991, 333: 15 July 1944.
11. Frank 1991, 281–82: 3 May 1944.
12. Frank 1991, 318–19: 13 June 1944.
13. Katz 1997.
14. Schama 1995.

~ Chapter One ~

THE WARSAW CEMETERY AND THE LIBERATION OF NATURE

I.

The trees are all around me. I could be in a forest, yet I can hear the sounds of traffic on Okopowa Street on the other side of the wall. Inside the Jewish Cemetery of Warsaw all is quiet. It is October 1995, fifty years after the end of the Second World War, and I have come to witness some of the remains of Jewish history in Eastern Europe, the landscape of the Holocaust. There is a light rain and fog. In the grayness of the day, the mist and the shadows prevent my eyes from seeing deep into the cemetery. What I can see are the trees and the underbrush, lush and green, growing up and over the scattered and crooked gravestones. One main walkway and a few paths that branch out from it have been cleared, so that visitors can view several hundred of the tombstones. Another open path leads to a clearing. It is a clearing of tombstones, not of trees, for it is the mass grave of the Jews who died in the Warsaw ghetto before the deportations to the Treblinka death camp began in July 1942. The mass grave takes the form of a meadow under a canopy of tree branches. Gravestones ring the meadow as a broken border fence, but the center of the clearing is covered with grass. Dozens of memorial candles flicker, remaining lit despite the dampness and the light rain. The beauty of this mass grave surprises and shocks me. Here is the physical incarnation of irony. This cemetery, a monument to the destructive hatred of the Nazi Holocaust, is extraordinarily beautiful. Filled with a vibrant, unchecked growth of trees and other vegetation, the cemetery demonstrates the power of nature to re-assert itself in the midst of human destruction and human evil.

The next day I travel to Lublin, near the Ukrainian border. This is a two-hour drive from Warsaw, through endless flat farmland where Polish farmers still use horses to plow the fields. It is harvest season, and the car slows occasionally to pass a truck piled high with sugar beets. Our destination is Majdanek, the Nazi death camp lying three kilometers

The Warsaw Cemetery and the Liberation of Nature

from the center of Lublin. Majdanek fills a treeless meadow stretching as far as the eye can see. I stand at the entrance gate and observe, about a thousand yards in the distance, the chimney of the crematorium.

Unlike Treblinka or the more famous Auschwitz-Birkenau, the camp at Majdanek was built near a major urban center; indeed Lublin would supply about five thousand of the victims murdered in the camp.[1] Majdanek was not hidden in the countryside. It is easy to imagine the smoke from the crematorium drifting into the heart of downtown Lublin. Likewise, it is hard to believe that the people of Lublin did not know what was happening at the camp. Lublin was the headquarters for Operation Reinhard, the plan to kill the entire Jewish population of the conquered land of Poland. Majdanek itself was first established as a slave labor camp in 1940, but its gas chambers began operating in November 1942. Approximately 360,000 people were killed at Majdanek—200,000 were Jews and the rest were non-Jewish Poles and Soviet prisoners of war. They died by the gas chamber, by shootings, and by overwork, disease, and malnutrition. In one day alone, November 3, 1943, 18,000 prisoners were shot and killed, their bodies piled into open ditches near the crematorium. Over 800,000 shoes were found at the camp when it was liberated in July 1944 by the advancing Russian army. Majdanek was the first of the Nazi death camps to be liberated, the first to be seen by the Allied forces and the Western media. Most importantly, because the camp was liberated so early in the last year of the war, the SS command structure had not yet developed a plan to deal with camps that fell into Allied control. Unlike the camps farther west that were liberated later, Majdanek was not destroyed by the retreating German forces. Although many of the wooden barracks buildings have deteriorated through natural decay, the camp as a whole exists today as it did in 1944, relatively intact.[2] It remains as a monument to human evil and destruction.

I stand in the small open courtyard a few dozen yards beyond the main entrance gate. On this spot the selections of arriving prisoners were made—who would live and work in the camp, who would be killed immediately. To my right is the gas chamber, a wood-shingled building, painted brown, with a sign that reads "Disinfection Bath" in German. Behind the gas chamber were the open pits for burning corpses, a supplement to the ovens of the crematorium building at the other end of

Chapter One

the camp. On my left is a row of barracks, used as storerooms and work areas when the camp was in operation. These unheated and dimly lit buildings now house museum exhibits. Beyond them is the main camp, divided into several sections or compounds. Each section consists of two rows of barracks facing a wide open parade ground. I enter the gate that permits entry through a double row of barbed wire and wooden fencing and walk through the parade ground of the first section of barracks, what was once the women's compound. I head out onto a road along the perimeter of the camp, a road that leads to the crematorium and the site of the November 1943 mass shooting. The camp is virtually empty of visitors. As in Warsaw the day before, there is a light rain and mist, and the autumnal air is cold, signaling the arrival of winter.

As I stand near the crematorium, overlooking the landscape of the concentration camp, my mind struggles to comprehend two opposing perceptions. The death factory of Majdanek is too beautiful. The green grass of the parade ground suggests a college campus, not a site of slave labor and mass executions. Is it possible to stand here in this grassy meadow and imagine the mud, the dirt, the smell—the unrelenting gray horror of the thousands of prisoners in their ill-fitting striped suits standing at roll calls? Is it possible to imagine the perpetually gray sky, filled with smoke and ash from the crematorium and the burning pits near the entrance of the camp? Perhaps it would be better to see Majdanek in the middle of the winter when one is not overwhelmed by the color of the green grass. As in the Warsaw cemetery the day before, nature prevents me from seeing, understanding, and feeling the true dimensions of the remnants of the evil that confronts me.

The experience of these two places—the cemetery and the death camp—raises questions for me about the healing power of nature in its relationship with human activity. And thinking of the healing power of nature in these historically unique situations leads me, in turn, to raise questions about both the ontological and the normative status of nature: what is nature, and why and how is it valuable? Can a study of the Holocaust reveal any truths about nature and the environmental crisis that surrounds us in the contemporary world? Can the study of nature and natural processes teach us anything about the evil of human genocide? Can the study of genocide teach us anything about the human-induced destruction of the natural world, what is sometimes called the

The Warsaw Cemetery and the Liberation of Nature

process of "ecocide," in an obvious attempt to equate it with genocide? These are not subjects that permit facile comparisons and analogies. Generally we study the Holocaust and the environmental crisis from different perspectives, with different attitudes and purposes. Holocaust studies and environmental philosophy are not generally thought to be compatible subjects for analysis and discussion. Yet the comparison may be helpful; indeed it may be full of profound meaning. Again, I return to a consideration of Anne Frank's tree. With this tree we can see how a natural entity can be a symbol of hope in a world that has become overwhelmingly evil, as it was for Anne. The evil that this tree confronts is the evil of domination. Perhaps the idea of domination can be used to link together an analysis of the Holocaust and the destruction of the natural world. Perhaps this comparison then can point us in the direction of developing a harmonious relationship with both the natural world and our fellow human beings.

II.

I want to emphasize the importance of my visit to the actual sites described above, and indeed, to those places I will describe in subsequent chapters. This book contains more than a philosophical argument. I could not have developed the ideas set forth in these pages through the typical philosophical methods of argument, analysis, example, and rebuttal. The lived experience of these places not only colors my ideas but also completely informs them. Indeed, this book is a written expression of my attempt to understand the physical experience of these Holocaust sites, to situate these experiences in the context of my philosophical thoughts about the meaning of the environmental crisis and the practice of human domination.

Why should I even try to connect these two areas of inquiry? Why think about the environmental crisis and the Holocaust in terms of one another? Is there a meaningful relationship between human ideas of the natural world and the concepts of domination and genocide? The Nazis thought so. As we shall see in much more detail in later chapters, the reconstruction and development of Polish farmland under scientific principles of management was one of the major goals of German settlement in the conquered lands east of Germany. Quoting from a contemporary record, architectural historian Robert Jan van Pelt describes

Chapter One

a trip through Poland in 1940 undertaken by Heinrich Himmler, the Reichskommissar for the resettlement of the German people, and, arguably, the second most powerful man in the Nazi hierarchy after Hitler himself. Himmler and his personal friend Henns Johst stand in a Polish field, holding the soil in their hands, and dream of the great agricultural and architectural projects to come: the re-creation of German farms and villages, the replanting of trees, shrubs, and hedgerows to protect the crops, and even the alteration of the climate by increasing dew and the formation of clouds.[3] As part of this plan to Germanize the landscape, there would have to be, of course, an "ethnic cleansing" of the region. The Polish people, both Gentile and Jewish, would have to be moved elsewhere or otherwise eliminated so that a German agricultural utopia could be developed. Fortunately for the realization of German policy goals, Himmler, as leader of the SS, was also in command of all operations that would produce this ethnic cleansing. And so we are introduced to the idea that the control of nature—in the re-development of the landscape, including the climate, to create a German agricultural homeland—was a central part of the Nazi plan. The domination of nature and humanity are linked together.

The domination of nature, of course, has long been a goal of Western civilization. It remains so, even today. As I have argued in my earlier work,[4] the primary goal of the Enlightenment project of the scientific understanding of the natural world is to control, manipulate, and modify natural processes for the increased satisfaction of human interests. Humans want to live in a world that is comfortable—or at least in a world that is not hostile to human happiness and survival. Thus the purpose of science and technology is to comprehend, predict, control, and modify the physical world in which we are embedded. This purpose is easy to understand when we view technological and industrial projects that use nature as a resource for economic development. Yet the irony is that the same purpose, human control, motivates much of environmentalist policy and practice.

As examples, let us consider the arguments of two writers on the theory and practice of environmental policy: Martin Krieger's call for artificial wilderness areas that will be pleasing to human visitors, and Chris Maser's plans for re-designing forests on the model of sustainable agriculture.[5] Maser is an environmentalist and Krieger is not; yet their

The Warsaw Cemetery and the Liberation of Nature

views on environmental policy are strikingly similar. Maser was once considered a spokesperson and leader of enlightened environmental forestry practices, but his goal is to manage forests in such a way as to maximize the wide variety of human interests in forest development: sustainable supplies of timber, human recreation, and spiritual and aesthetic satisfaction. Krieger is a public policy analyst interested in the promotion of social justice. His goal is to develop an environmental policy consistent with the maximization of human economic, social, and political benefits. Thus he argues that education and advertising can re-order public priorities, so that the environments that people want and use will be those available at the lowest cost. Natural environments need not be preserved if artificial ones can produce more human happiness at a lower cost.

What ties together views such as Krieger's and Maser's is their thoroughgoing anthropocentrism: i.e., human interests, satisfactions, goods, and happiness are the central and fundamental goals of public policy and human action. This anthropocentrism is, again, not surprising. Since the Enlightenment, at least, human concerns—rather than the interests of God—have been the central focus of almost all progressive human activities, projects, and social movements. The institutions of human civilization are planned, organized, and structured to improve the lives of human beings. Although methods may differ, and the set of people that is the primary object of this concern for improvement may differ, the central anthropocentric focus is consistent regardless of ideology or social position or political power. Humanity is in the business of creating and maximizing human good.

Anthropocentrism as a worldview easily leads to policies and practices of the domination of nature, even when the domination is not articulated. Indeed, in much of progressive environmental policy, the domination of nature by and for human interests is not even recognized or understood. Environmental policy often conceives of the natural world as a nonhuman "other" to be controlled, manipulated, modified, or destroyed in the pursuit of some human good. As a nonhuman other, nature is understood as merely a resource for the development and maximization of human interests; as a nonhuman other, nature has no valid interests or good of its own. Consider the most influential and popular environmental idea of the last three decades: sustainable development.

Chapter One

Since the 1987 Brundtland report, which defined sustainable development as "development that meets the needs of the present without compromising the ability of future generations to meet their own needs," the policies and activities of governments, industries, NGOs, communities, and even individuals have aspired to follow, or at least pay lip service to, this basic idea. But sustainable development is an idea that is highly anthropocentric in character, and it leads to policies and actions that privilege human concerns over the wellbeing of the natural world. The basic concept in the Brundtland definition is that future generations of human beings are to be able to meet their needs, to satisfy their interests. So at a basic level sustainable development is an economic idea that foresees continual economic growth (i.e., development) through time, so that present human populations as well as future human populations maintain an acceptable economic lifestyle. Yet it is also supposed to consider the overall quality of life for present-day and future humans beyond mere economic wellbeing. The sustainable economic development is supposed to be balanced against environmental degradation, for the problems created by the destruction of nature (e.g., pollution, global warming, etc.) will have serious negative impacts on the human quality of life. Thus, sustainable development calls for the furtherance of human welfare and economic wealth but at the same time the conservation of nature and natural resources. But the reason or motivation for maintaining a healthy and functioning natural environment, at least as a long-term resource base, is the continued production of goods and benefits for human beings. Sustainable development is thus an expression of an anthropocentric worldview, for its central focus is the welfare of human beings, now and in the future. Nature is merely the nonhuman other that is used to produce this human wellbeing.

Another environmental policy, ecological restoration, can also be considered to be an expression of anthropocentrism. A thoroughgoing analysis and criticism of ecological restoration is one of the major themes of this book, and I will develop this argument more fully below in this chapter and in Chapter Three, but for now we should note that the restoration of degraded ecosystems to a semblance of their original states is a policy that is permeated with anthropocentric ideology. Under this policy, natural ecosystems that have been harmed by human activity are restored to a state that is more pleasing to the current human popula-

The Warsaw Cemetery and the Liberation of Nature

tion. A marsh that had been landfilled is re-flooded to restore wetland acreage; strip-mined hills are replanted to create flowering meadows; acres of farmland are subjected to a controlled burn and replanting with wildflowers and shrubs to re-create the oak savanna of the pre-European American plains. We humans thus achieve three simultaneous goals: we create an improved ecosystem or natural area that is more in line with our current interests and desires; we relieve our guilt for the earlier destruction of natural systems by creating a functional replacement; and we demonstrate our human power—the power of science and technology—over the natural world.[6]

But the domination of nonhuman nature need not be the only result of an anthropocentric worldview. The ideology of anthropocentric domination may also extend to the oppression of other human beings, those conceived as a philosophical "other," as nonhuman or subhuman. As C. S. Lewis wrote seventy years ago, at the end of the Second World War, "what we call man's power over Nature turns out to be a power exercised by some men over other men with Nature as its instrument." The reason that this exercise of power is considered to be justifiable is that the subordinate people are not considered to be human beings: "they are not men at all; they are artefacts."[7] It is here that we can see the connection between the domination of nature that is manifested in the environmental crisis of the contemporary world and the domination of humanity that was manifested through the genocidal policies of the Holocaust. Anthropocentrism does not convert automatically into a thoroughgoing humanism, wherein all humans are treated as equally worthwhile. As we know from history, for example, the idea of human slavery has been justified from at least the time of the ancient Greeks (and probably long before into prehistory) by designating the slave class as less than human (as in Aristotle's *Politics*[8]). In the twentieth century, the evaluation of other people as subhuman finds its clearest expression in the Nazi ideology concerning the Jews (and the Slavs and Romany), but we find its echoes in the contemporary world, be it the ethnic civil war in the former Yugoslavia, the genocide in Rwanda (where the Tutsis were described as "cockroaches"), or the hatred of the Palestinians by some extreme right-wing Israelis. We generally recognize that any form of ethnocentrism or racism can easily lead to prejudice, oppression, and domination regarding a denigrated class, but the same is true of anthropocentrism in general. The oppressed class—be it a specific race or religious group, or even animals or natural entities—is simply denied admittance to the elite center of value-laden beings.[9] From within anthropocentrism, only humans have value and only human interests and goods need to be pursued. But who or what counts as a human is a question that cannot be answered from within anthropocentrism; this question requires an external standpoint to determine the normative and ontological status of any entity or set of entities. And the answer to this question will determine the likely extent of the practice of domination.

We have thus arrived at a provisional answer to the question that began this section. Environmental philosophy and Holocaust studies are not only compatible but they are also mutually reinforcing lines of inquiry. The ideas of anthropocentrism and domination tie together a study of the Holocaust and the contemporary environmental crisis. Whether we consider genocide, the destruction of a human people, or ecocide, the destruction of natural systems and entities, we find the justification that the victims are less than human, that they exist outside the primary circle of value.

III.

The resurgence of trees in the Warsaw cemetery and the lush green grass of the meadow at the Majdanek death camp serve as catalysts for rethinking the relationships among nature, humanity, and the practice of domination. In these places, one can describe metaphorically the processes of nature as a kind of healing, a soothing of the wounds wrought by the evil of the Holocaust. Does nature, over time, make everything better? Can we say that dominated and oppressed entities are saved—redeemed—by the ordinary processes of the natural world as they correct the evil that humans perpetrate? Does nature have this power? And if it does, what are the implications for the way in which humanity acts in relationship to the natural world?

First, we should note an objection to this entire line of analysis. One might argue that in thinking of nature as having a redeeming power over human evils, we are, in part, treating nature as if it possessed a kind of intentional activity. But nature is not a rational subject. Nature makes no decisions, rational or otherwise. To think that nature makes acts either rationally or irrationally is to make what philosophers call a

The Warsaw Cemetery and the Liberation of Nature

"category-mistake"—it is to apply the wrong concept to a situation or entity. If the lush vegetation hides the horrors of Majdanek this is not the result of any natural plan, but merely the effects of natural processes in their normal operations. According to this objection, we should be wary of anthropomorphizing natural processes, of being misled by metaphor and analogy.

This objection serves as an important warning to the analysis that follows in this book. Nature has no intentions—and no other thoughts, desires, wants, or needs. Nevertheless, I do believe—as I noted briefly in my Preface—that nature can be considered to be analogous to a human subject. Human actions can benefit or harm natural processes in ways similar to the benefits and harms produced for other humans, for human institutions, and for nonhuman living beings. Moreover, nature does act in predictable ways similar to a thinking being. As Colin Duncan has claimed, "While Nature is certainly not a person ... it does have some of the attributes of a Hegelian subject. It can be both victim-like and agent-like."[10] Most important for my thesis, we can consider nature as the subject of an ongoing history that can be interfered with or destroyed by human action. From the perspective of normative axiology (value theory), nature develops in ways similar to human subjects—the continuous processes of nature produce good and bad consequences for itself and for other entities. Morally and axiologically, then, nature can be considered to be equivalent to a subject. Without anthropomorphizing nature—without attributing to it the emotions, feelings, desires, and rational will of human subjects—we can understand that it is not merely a passive object to be manipulated and used by humanity.[11]

Nature, in fact, acts upon human beings, human institutions, and the products of human culture in powerful ways. What we call natural disasters, such as tsunamis, earthquakes, and floods, are the prime examples of events in which natural forces impact humanity. But ordinary weather, small variations in climate, disease organisms, the migration of species, and even the rotation of the earth are also activities of nature—natural processes—that affect human life. Elsewhere I have categorized this type of natural activity as nature's imperialism over humanity, for it has a parallel structure to the basic kind of human imperialism over other humans, as well as to the human imperialism over nature. Imperialism is a form of domination, in which one entity uses, takes advantage of, controls, or otherwise exerts force over another. If we consider nature as both a possible subject and object of imperialism, then we can think of nature as exerting its power—attempting to dominate—humanity, just as we can think of humanity attempting to dominate nature.[12]

But my experiences in the Warsaw cemetery and at Majdanek suggest that nature's domination in these places is benign. Nature appears to heal the wounds of human atrocities, to cover the scarred remains of human evil. Nature here does not appear to exert the oppression of an imperialist. Nature appears to provide the balm to restore the health and goodness of a world disrupted and harmed by the intentional acts of evil human beings. Nature's domination—its resurgence in these realms of human atrocities—serves as the corrective to the effects of human domination, in this case to the oppression and genocide of Eastern European Jewry. Is this an appropriate way to interpret the experiences of these places?

I think not. One objection to viewing nature as a benign healer of human-induced wounds is that such a perspective on nature is yet another expression of an anthropocentric worldview. Rather than use nature as a physical resource for economic purposes, we are here using nature as an emotional resource, to make us feel better about the horrors of human destruction.[13] We are blinded to the fact that natural processes develop independently from human projects; nature follows its own logic. A forest re-grows after a burn caused by a lightening strike; a tidal marsh is rejuvenated after a storm surge. Nature can and does create new life and new beauty. Yet none of these natural activities are properly described as a "healing," since that characterization implies human intentionality. So the desire to see nature as a healer demonstrates how pervasive is the anthropocentric perspective. We humans seem incapable of viewing the natural world on its own terms, free of the categories and purposes of human life and human institutions.

Even more importantly, the question arises whether or not nature *can* heal these wounds of human oppression. Consider the reverse process, the human attempt to heal the wounds that have been wrought upon nature. We often tend to clean up natural areas polluted or damaged by human activity, such as the Gulf Coast harmed by the BP oil drilling disaster of the spring of 2010. But we also attempt to improve natural areas dramatically altered by natural events, such as a forest damaged

The Warsaw Cemetery and the Liberation of Nature

by a massive brush fire, or a beach suffering severe natural erosion. In most of these cases, human science and technology are capable of making a significant change in the appearance and processes of the natural area. Forests can be replanted, oil is removed from the surface of bays and estuaries, sand and dune vegetation replenish a beach. But are these activities the healing of nature? Has human activity—science and technology—restored nature to a healthy state?

For over twenty-five years I have written essays and book chapters arguing that the answer to this question is a resounding "no."[14] In Chapter Three, I will delve more deeply into this critique of ecological restoration, answering my critics and pushing the argument into new directions. But for now all that is required is a brief summary argument highlighting the main points of my position. In general, I believe that when humans intentionally modify a natural area they create an artifact, a product of human labor and design. This restored natural area may resemble a wild and unmodified natural system, but it is, in actuality, a product of human thought, the result of human desires and interests. All humanly created artifacts are manifestations of human interests—from computer screens to rice pudding. An ecosystem restored by human activity may appear to be in a different category—it may appear to be an autonomous living system uncontrolled by human thought and action—but it nonetheless exhibits characteristics of human design and intentionality: it is created to meet human interests, to satisfy human desires, and to maximize human good.

Consider again my examples of human attempts to heal damaged natural areas. A forest is replanted to correct the damage of a fire because humans want the benefits of the forest—whether these are timber, a habitat for wildlife, or protection of a watershed. The replanting of the forest by humans is different from a natural re-growth of the forest vegetation, which would take much longer and would likely include different individual plants and species. The forest is replanted because humans want the beneficial results of the mature forest in a shorter time, or with a prescribed population of specific trees and other vegetation. Similarly, the eroded beach is replenished—perhaps with sand pumped from the ocean floor several miles offshore—because the human community does not want to live with the natural status of the beach. The eroded waterfront threatens the oceanfront homes and recreational beaches. Humanity prefers to restore the human benefits of a fully protected beach. The restored beach will resemble the original, but it will be the product of human technology, a humanly designed artifact for the promotion of human interests.

After these actions of human restoration and modification, what emerges is a nature with a different character than the original. This difference is an ontological difference, that is, a change in the essential qualities of the restored area. A beach that has sand replenished by human technology possesses a different essence than a beach created by natural forces such as wind and tides. A savanna replanted from wildflower seeds and weeds that were collected by human hands has a different essence than grassland that develops on its own. The source, the genesis, of these new areas is different—human-made, technological, artificial. The restored nature is not really nature at all.

A nature "healed" by human action is thus not nature. As an artifact, it is designed to meet human purposes and needs—perhaps even the need for areas that look like a pristine, untouched nature. In using our scientific knowledge and technological power to "restore" natural areas, we actually practice another form of domination. We use our knowledge and power to mold the natural world into a shape that is more amenable to our desires. We oppress the natural processes that function independently of human power; we prevent the autonomous development of the natural world. To believe that we heal or restore the natural world by the exercise of our scientific knowledge and technological power is, at best, a self-deception and, at worst, a rationalization for the continued degradation of nature—for if we are confident that we can heal whatever damage we inflict, we will face no limits to our actions regarding the natural world.

This conclusion has serious implications for the idea with which we began, the idea that nature, conversely, can repair human destruction, that nature can somehow heal the evil that humans perpetuate on the earth. Just as a landscape restored by humans has a different causal history than the original natural system, the re-emergence of nature in a place of human genocide and destruction is based on a series of human events that cannot be erased. The natural vegetation that covers the mass grave in the Warsaw cemetery is not the same as the vegetation that would have grown there if the mass grave had never been dug. The grass and

The Warsaw Cemetery and the Liberation of Nature

trees in the cemetery have a different cause, a different history, that is inextricably linked to the history of the Holocaust. The grassy field in the Majdanek parade ground does not cover and heal the mud and desolation of the death camp—it rather grows from the dirt and ashes of the site's victims. For anyone who has an understanding of the Holocaust, of the innumerable evils heaped upon an oppressed people by the Nazi regime, the richness of nature cannot obliterate nor heal the horror.

IV.

Rather than think of nature as a force that can heal humanly created wounds or that can overcome the evils of humanly created oppression and domination, we ought to think of nature as being in need of a liberation of its own. The liberation of nature would seem to be a necessary implication from the idea, introduced above, that nature can be considered to be analogous to a human subject. A subject, after all, can be free or oppressed. Also important would be the connection to the guiding image of this book—Anne Frank's tree—for in her diary the image of a nature independent of the forces of human evil and destruction is the foundation of her hope for a better world. But what can the "liberation of nature" mean?

In *Counterrevolution and Revolt*, Herbert Marcuse declared that "nature, too, awaits the revolution!"[15] Nature, in other words, has a possible future free of human domination. Although I do not plan to analyze the work of Marcuse or other critical theorists, I will use this remark as a starting point to consider the meaning of the idea that nature can be liberated, that it can be released from human domination. I choose this pithy remark of Marcuse as a starting point because the field of mainstream environmental ethics has said surprisingly very little about the domination and possible liberation of nature. One classic title, William Leiss's *The Domination of Nature*, was a study of Francis Bacon and not a treatise on environmental ethics.[16] Perhaps the only sustained discussion of the concepts of liberation and domination in the field of environmental ethics has been in the work of ecofeminist philosophers. At least since 1980, when Carolyn Merchant published *The Death of Nature*, ecofeminist philosophy has emphasized as its primary theme the connection between the domination of women and the domination of nature.[17]

Chapter One

But as early as 1977, John Rodman, with perhaps an ironic eye towards Marcuse's essay, published "The Liberation of Nature?," a critique of both Peter Singer's idea of animal liberation and Christopher Stone's proposal for the establishment of legal rights of natural entities as models of a new environmental consciousness.[18] If nature were to be truly liberated, Rodman argued, we would have to do better than extending utilitarianism to the animal kingdom (following Singer) or granting rights as convenient legal fictions to nonhuman natural objects (following Stone). Taking as his symbolic act of defiance the freeing of captive dolphins, Rodman insisted that we must resist the technological monoculture that is rapidly enveloping the contemporary world.

My use of the concept of domination—and the idea that I take to be its opposite, autonomy—has so far been fundamental to the argument of this book, yet I have used these ideas uncritically. I have been reluctant to enter into any serious metaphysical debates about the meaning of human nature or the nature of nature itself. Yet when I claim that nature should be treated as analogous to a human subject, that nature needs its own revolution of liberation, or that a nature free of human domination should be the primary goal of human activity regarding the natural environment, I open the door for critical questioning about the metaphysical foundations of my position. What is this nature that is analogous to a human subject? What exactly is this autonomous entity that needs to be liberated from the chains of human domination?

The precise locus of my problem concerns the existence and description of nature in itself, the nature of nature. I have been inspired by the vision that Anne Frank had about her chestnut tree, that the autonomy and self-development of nature is to be respected and used as a motivation for human action. For me, in other words, nature is to be treated as a moral subject. But if this vision of nature as an autonomous moral subject is to have any meaning and practical force, then we need some sense of what nature is, in itself, outside the domain of human activity. The problem for a critical philosophical analysis is that nature is only known through human activity, and even more problematic, nature is continually modified by human activity. Thus both epistemologically and ontologically, nature in itself is "our" nature, the nature constructed by human thought and praxis. Can there be a nature that exists in itself, independent of human life, thought, and action? According to Steven

The Warsaw Cemetery and the Liberation of Nature

Vogel, in his book *Against Nature*, the problem of nature in itself is also the problem for critical theorists such as Marcuse and Jürgen Habermas—"how to reconcile an account of knowledge as active and social ... with the 'materialist' commitment to a nature independent of the human."[19] But this problem is more than a problem for critical theory—it is a problem for any philosophy concerned with the human relationship to the natural world. Any environmental philosophy that deals with a robust nonanthropocentrism must have a clear sense of what nonhuman nature is. Any account of environmental ethics that extends moral consideration beyond the boundaries of the human species would seem to require some idea of what nature and natural entities are in themselves, free of human influence and control. We need to know what is good for nature in itself in order to act for this good.

The problem is that we know and understand nature through human categories. For example, we use human conceptions of good to evaluate the processes of nature, the flourishing of natural entities and systems. The human interest in nature is the factor that focuses our perceptions and understanding of the natural world. If nature is understood in this way, it does not appear that it could ever be free of human domination, for the basic domination is epistemological: nature is only known through human thought. For the operation of a nonanthropocentric environmental ethic, or for the existence of an ideal nature independent of humanity that can be used as a source of hope and resistance, we seem to require an idea of a nature that is autonomous, a nature that is analogous to a human subject, so that we can preserve and promote the interests of this nature in itself. But to think of a free and autonomous nature, it seems, means that we must think of a nature that is completely free of human influence, to think of nature in itself, a nature that is outside of all human categories of thought.

But can we know what nature is in itself? Given our post-Kantian understanding of human thought, it seems unreasonable to think that we can know nature in itself, or what Kant called *an sich*. But is knowledge of what Kant termed the noumenal world of nature really required for the development of a nonanthropocentric environmental ethics? Do we need to know the fundamental ontological reality of nature to recognize its existence independent from human institutions and concerns? Perhaps I have approached this problem in the wrong way. Perhaps there is no real need for a metaphysical examination of nature as such. Here I want to suggest a pragmatic response to these questions: perhaps we can avoid metaphysical speculations about the nature of nature. Perhaps we can "make do" with the concepts and practices that we have at our disposal as practical moral philosophers.

Let me offer a tentative pragmatic solution to the problem of nature in itself. Is there a nature outside the knowledge and activity of human society that can be a subject unto itself? Is there a nature that can be liberated from human domination? For an answer, let us compare the problem of the liberation of nature with that of the liberation of humans. Given the limits of our epistemology, we do not really know what humans are in themselves either. The Kantian analysis of the knowledge of physical nature—that we can know only the world of phenomena, for the noumenal world is filtered or structured by our human categories of thought—applies to humans in their physical being as well. I do not know other human beings, nor even myself, outside of socially constructed categories. All of my relationships with all individual human beings and all human groups and institutions are mediated by cultural constructs and social roles. And yet in my relationships with other humans and human institutions I can meaningfully strive to end oppression and domination, to aid other human beings in achieving liberation, freedom, and autonomy. I do not require an idea of a human being in itself for a meaningful liberatory praxis.

So what does liberation mean? It does not mean the elimination of all social constructs and categories. A human being does not become liberated when he or she transcends all social and cultural roles, duties, and obligations. Even if this kind of transcendence were possible—which it cannot be—what could it possibly mean? A pure human essence existing outside of all human history, free of all the rules of human social life? The prehistorical natural or biological human? Although such an abstract ideal may have a place in the conceptual analysis of the meaning of human life, it surely plays no part in our daily practice of working towards the liberation of individual humans and human institutions.

Regarding the liberation of humans, then, my point is this: we do not need an idea of an ideal human nature in order to understand practices of liberation and domination that we encounter in the everyday world. There are, of course, difficult cases. As a parent, for example, I

The Warsaw Cemetery and the Liberation of Nature

have long been fascinated by the boundaries of education, socialization, indoctrination, and oppression in my relationships with my growing children. But the existence of gray areas and marginal cases does not in the least prevent me from recognizing the real oppression of children by their parents. My parenting, I hope, is always guided by both an understanding of the appropriate uses, abuses, and limitations of my authority, and a rather nebulous idea of a maturing autonomous human being in contemporary culture—the characteristics that I hope will develop in my children. Similarly, in the broader social and political sphere, we do not require an idea of an ideal human nature in order to oppose (for example) slave-labor practices, various forms of racial, gender, and religious discrimination, economic injustice, and imperialism. Our social context informs our decisions. What we mean by human liberation is embedded within our social categories, which may, of course, change as society itself becomes more (or less) liberated. So human liberation is the development of specific positive freedom- and life-enhancing roles, not the elimination of all social constraints, commitments, constructs, and categories. Although there will continue to be difficult cases, our ethics and our social praxis are enough. We need not turn to metaphysical speculation on the essence of humanity to give a robust normative content to our activities regarding human liberation.

Why is it not the same for our relationship with nature? Why do we need an idea of a nature in itself, outside of all human categories of knowledge and action, to give content to a robust nonanthropocentrism or to provide the basis for an idea of a nature free from human domination and evil? Surely our practical activities in their interaction with nature are enough to give us a sense of what is right and wrong—as it was for Anne Frank as she contemplated the meaning of her chestnut tree. Do I really need an idea of nature in itself, the nature of nature, to know that clear cutting a forest is a form of domination, an injury to the autonomous development of the forest ecosystem? Do I really need an idea of a nature unmediated by human categories of thought and action to know that damming a free-flowing river interferes with the spontaneous movement of natural processes—or that the BP oil spill was harmful to the ecosystem of the Gulf of Mexico? Without denying that there will be difficult cases, it seems clear that we know what is involved in the domination (and thus, the liberation) of nature. Environmentalist practice informs our decisions; we have no need for metaphysical inquiries into the nature of nature as such.

To return to Marcuse's claim: nature also awaits the revolution, its liberation. Can we give a concrete example of what this means before we re-enter the world of Anne Frank and the Nazi domination of nature and humanity? What is the liberation of nature? What does the autonomy of nature look like? Consider an environmental issue dear to my heart: the ethics of beach preservation and sand replenishment projects on Fire Island, a barrier beach off the coast of Long Island in the Atlantic Ocean, where I have a summer home. Fire Island is an interesting case because it is a hybrid environment. The island is thirty-two miles long and at its thickest about a half-mile wide—it is, essentially, a long sandbar. Although there is no large-scale commercial development, some sections of the island are densely populated with individual homes on small lots. But most of the island remains undeveloped. There is a unique wilderness area in the central part of the island—the Sunken Forest—and the island is home to several threatened and endangered species of plants and birds. In 1964 the Federal Government purchased the island and made it part of the National Seashore, roughly equivalent to a national park.

As with all barrier beaches on the Eastern coast of the United States, Fire Island suffers from erosion. Individual homes, recreational beaches, and the wilderness areas are threatened by the loss—the movement—of sand. Whether or not a policy of beach replenishment should be undertaken is a question that raises interesting issues in technology, economics, social justice, and environmental ethics. I do not address those issues here.[20] In this chapter I am only concerned with the idea of the autonomy of nature. Can we look at the problem of beach erosion and the environmental policy of beach replenishment from the perspective of the liberation of nature?

Let us begin with the assumption that to liberate nature in this case, to permit the autonomy of natural processes, we would adopt a "hands-off" policy regarding beach erosion and replenishment. Rather than trying to mold and manipulate the beach environment, we would simply leave it alone—thus permitting both the natural erosion (and sometimes, the natural accretion) of sand to continue. But Fire Island is not a natural environment—as I mentioned above, it is a hybrid area of wilderness, relatively undisturbed beaches, and single-family homes.

The Warsaw Cemetery and the Liberation of Nature

There are concrete and wooden walkways, a few unpaved sand and dirt roads, extensive bulkheading, and numerous boat channels and harbors. It is as much a built and human environment as a natural or wild one, and this human presence has a significant impact on the natural movement of the sand that comprises the beach and the island. The human presence makes the entire idea of the autonomy of natural processes rather suspect. Only if we were to systematically eliminate all human-built structures and modifications to the shoreline could we begin to approximate a natural environment (although such a situation would resemble an ecological restoration project, and thus probably not meet my idea of a natural system). Only on an island with no human structures or human presence could the idea of the liberation of nature make sense.

In the real world the systematic elimination of all human structures on the island is not going to happen. So let us undertake a philosophical thought experiment. Imagine an island identical to Fire Island—thirty-two miles long, central wilderness area, threatened and endangered species—but without a permanent human presence. No houses, no harbors, no boat channels, no sidewalks or roadways, and no bulkheading. On this imaginary island, what would the liberation of nature be like? Clearly, it would be *the continuation of the freedom from human impacts*. The autonomy of nature would be the unfolding of natural processes on the island—and the island's interactions with the ocean and bays—without the interference of humans, without the human development and alteration of the land. Nature would develop in its own way, not subject to the designs, plans, or projects of humanity. And to say that nature would develop in its own way does not imply that nature itself has a plan, a goal, or a *telos*. Rather, we are simply eliminating the dominating tendencies of human plans, human intentionality and design.

This imaginary island thought experiment shows that we do not need a positive conception of nature as such to understand the idea of the liberation and autonomy of nature. We do not need to know a nature outside of all human categories—indeed, the idea of nature that we have on this imaginary island is an idea constructed by our science: it is a nature that we understand through human categories. But this does not make it any less autonomous. As long as it is not being molded and transformed by human impacts it is a free and liberated nature. It may not be free of human domination in a metaphysical or epistemological sense, but in the realm of pragmatic environmental policy, it surely is.

Can the liberation of nature on this imaginary island help us in understanding and determining environmental policy on a real island, say the real Fire Island, with its complex and hybrid interacting human and natural ecosystems? Can it help us understand the process of domination and liberation in the world of Anne Frank, in the plans of the Nazi regime to oppress and dominate the landscapes and peoples of Eastern Europe? Let us first return to Fire Island, where I live, and where I need to know what environmental policies are morally justified. My argument and thought experiment show that even in hybrid environments we ought to lean towards leaving nature alone, we ought to minimize human impacts that affect the course of natural processes. In most cases, the mere absence of human domination will result in the liberation and autonomous development of nature. In actual policy decisions then, when we have a choice, we should choose the least intrusive, and hence least oppressive, policy of action. On Fire Island, for example, if we wish to protect the recreational beaches, the wilderness areas, and the endangered species, we ought to preserve the beach by a process of sand nourishment, using snow fencing to catch the windblown sand and planting dune vegetation to hold it in place. We should not build permanent structures such as rock jetties and sea walls. Of course, a full-scale policy discussion would require a much more detailed description of the specific facts of the concrete situation, and this is not the place for that discussion.[21] My philosophical point about the formation of policy is merely this: we can make decisions about the autonomy of nature without plumbing the metaphysical depths of nature in itself. Thus it makes perfect sense to speak of the liberation of nature, to think of nature as analogous to a human subject, and to believe in the existence of a nature that is independent of human domination.

V.

So what we see in the Warsaw cemetery and the Majdanek death camp is an example of the independent nature idealized by Anne Frank when she contemplated the meaning of her tree. Here nature is liberated, free to pursue its own course after the evil, destruction, and degradation produced by human history. We can even see in these

The Warsaw Cemetery and the Liberation of Nature

Holocaust sites another example of the imperialism of this nature free of human domination. Nature here acts—without an intention or design—to erase the remnants of human evil. To speak in metaphor, nature imposes its vision of the world on its human interpreters. But nature's vision is not our vision, and in these places it does not express the essence of our experience. Just as the human restoration of a degraded ecosystem turns a natural area into an artifact, nature's restoration of a site of human destruction alters the character of the site. This is why, as I argued in section III above, we cannot view the action of nature here as a kind of healing. Although the beauty of the trees in the cemetery cannot be denied, the meaning and value of the cemetery lies not in the re-emergent trees but in the historical significance of the Nazi plan to kill the Jews of Eastern Europe.

Nature's re-emergence at these Holocaust sites is, from the point of view of nature, a process of liberation, but from the point of view of humanity, it is an example of domination: the domination of meaning. Nature slowly exerts its power over the free development of human ideas, human memory, and human history. The actions of nature seem to attempt the eradication of the human meaning of these places. Now it may seem strange to think of the liberating processes of a free nature as a form of domination, but it is clear that one entity's acts of liberation can be seen as an act of domination over another entity. Consider Holocaust survivor Primo Levi's description of his liberation from Auschwitz. He recounts the series of baths that he and the other prisoners were given by the Allies: "it was easy to perceive behind the concrete and literal aspect a great symbolic shadow, the unconscious desire of the new authorities, who absorbed us in turn within their own sphere, to strip us of the vestiges of our former life, to make us new men consistent with their own models, to impose their brand upon us."[22]

But Levi also compares these baths of liberation with the "devilish-sacral" or "black-mass" bath given by the Nazis as he entered the universe of the concentration camps. All of these baths serve as symbols of domination—the molding of human beings into artifacts appropriate for their current situation: free man or prisoner. The cleansing of liberation is thus comparable to the oppression of imprisonment, for both actions deny the autonomy of the free human subject. What can be considered to be a kind of healing—the liberation of Holocaust survivors or the emergence of trees over a mass grave—can be an expression of domination, if it modifies or destroys the meaning and the freedom of the original entity.

To understand the multiplicity of the forms of domination and liberation is the first step toward developing a comprehensive ethic for evaluating human activity in relationship to both the natural environment and the human community. We must resist the practice of human domination in all of its forms. We must act so as to preserve the free and autonomous development of human individuals, communities, and natural systems. We must understand the moral limits of our power to control nature and our fellow human beings.

Marcuse believed that after the revolution, not only would nature be liberated, but humanity would create a new non-dominating science, founded on a new sensibility of passivity, receptiveness, and openness. The new science would involve "the ability to see things in their own right, to experience the joy enclosed in them, the erotic energy of nature."[23] I do not know if any of this is possible. Can we see nature in its own right, independent of human categories of thought? Is this not what Anne Frank thought about her tree, that it was somehow a symbol of the peace and healing that would envelop humanity? I am reminded of the last verse of the *kaddish*, the prayer that closes almost all Jewish services, and also serves as the prayer of mourning for the dead. This verse is a call for the healing power of peace. *Osay shalom bimromov hoo ya-ahsay shalom, olaynoo v'al kol yisroayl*—"May He who establishes peace in the heavens, grant peace unto us and unto all Israel." In viewing the Warsaw cemetery and the Majdanek death camp, I was moved by the hope, as was Anne Frank, that nature could be the agent that establishes peace. But nature alone cannot accomplish this. If there is a God, He works through human decisions. Only humans can understand the meaning and history of evil. Only humans who understand the need to control our power can halt the practice of domination, can halt the destruction of people and the natural environment. It is only through human actions that peace can be restored to our planet and our civilization. And so we turn in the next chapter, to a consideration of human history.

The Warsaw Cemetery and the Liberation of Nature

Notes to Chapter One

1. United States Holocaust Memorial Museum 1996, 100.
2. For a general discussion of Majdanek and the overall history of the Holocaust, see Yahil 1990, especially 362–63. See also Gilbert 1985, and the classic Holocaust history, Hilberg 2003 (3rd edition). Hilberg notes the quick evacuation of Majdanek (1045–46) and Gilbert cites Hitler's disgust that the camp was not destroyed (711).
3. van Pelt 1994, 101–03.
4. Katz 1992b; 1993; 1995; and 1997.
5. I have examined each of these thinkers in the past and will merely review, in brief, my analysis here. See Krieger 1973; Maser 1988; Katz 1979; 1991; 1992a; and 1997.
6. See Katz 1991; 1992a; 1992b; 1993; 1995; 1996b; 1997; 2000; 2002; 2009; and 2012.
7. Lewis [1947] 1983, 143, 146.
8. See Book I, paragraphs 4–5.
9. See, for a prominent example of this line of thinking, the seminal work by Peter Singer, *Animal Liberation*, 1975.
10. Duncan 1991, 8.
11. See Katz 1995 and Rolston 1988, esp. 342–54.
12. Katz 1995.
13. I am indebted to Avner de-Shalit for bringing this argument to my attention.
14. See Katz 1991; 1992a; 1992b; 1993; 1995; 1996b; 1997; 2000; 2002; 2009; and 2012.
15. Marcuse 1972, 74.
16. Leiss 1974.
17. Merchant 1980; Warren 1990.
18. Rodman 1977; Singer 1975 and Stone 1974.
19. Vogel 1996, 141.
20. See Katz 1999.
21. The example of snow-fencing to capture sand as a method of beach replenishment will return several times in the course of this book. Let me say here that the reason why a snow-fence is different in kind from a stone and rock sea wall is the physical fact of the permanence of the sea wall. A flimsy wooden slatted snow-fence does nothing to change the essential character of the physical environment; it is a mere passive and temporary object to collect sand blown by the wind. This is entirely different from a stone wall that is meant to stay in place for a long period of time (decades, if not more) and that has permanent effects on the processes of the natural system. But note that there will be appropriate places to build sea walls; namely, in environments and places that are already highly artifactual, such as a seaport. My thanks to Andrew Brennan for this last point.
22. Levi 1987, 8.
23. Marcuse 1972, 74.

~ Chapter Two ~

THOUGHTS ON THE HOLOCAUST IN THE SPANISH SYNAGOGUE OF VENICE: HUMAN HISTORY, TECHNOLOGY, AND DOMINATION

I.

I am sitting in the Spanish Synagogue in Venice, during the morning service on the Jewish holiday of Simchas Torah. I sit on a plain wooden bench—it is dark wood, beautiful and soft in its age—with my back against the southern wall, as I listen to several members of the congregation read from the Torah scroll. The *bimah*, the pulpit, is on the western wall, across the room from the ark, and all the pews run east to west, so that one sits facing the center aisle between the *bimah* and the ark. These two focal points are perfectly balanced, and although the *bimah* is raised above the floor and flanked by two marble columns in the Corinthian style, it seems accessible, open, and inviting. The golden ark opposite is framed in a marble arch, and above the arch is a painted starry sky in blue and gold. Except for my bench in the last row against the wall, all the pews have little wooden desks, so that one can appreciate the fact that the synagogue is called in Italian a *scuola*, a school—in Yiddish, a *schule*. Across the center aisle is a trellised screen about five feet high, shielding the eyes of the men from the women who sit on the north side of the synagogue. Somewhere on that side of the synagogue is my wife. I look at my watch and hope that she is not bored. She knows much less about Jewish rituals than I, and a strange service in a foreign country might initially be intriguing, but after a while it may become tiresome. She cannot even pass the time by skimming through the prayer book, for it is written in Hebrew and Italian.

I, however, am not the least bit bored—I am enchanted by the entire spectacle. And I am quite pleased with myself for having managed to get into the synagogue, a process that involved some amateur con artistry. I had been to Venice many times over the previous years,

Thoughts on the Holocaust

and had always made a point of visiting the Ghetto section of the city with its five extant synagogues. The word *ghetto* means *foundry* in Italian, and it is likely that the origin of the term as applied to an isolated and restricted living community comes from the history of Venice, when the Jews were forced to re-locate to the two small islands of the old and new foundries in the early sixteenth century. Here they built at least eight synagogues, five of which are now restored and operated by the Hebrew Museum of Venice. A tourist can only enter the synagogues on an official tour of the Museum, and during these tours only three of the five synagogues are open. In the past I had seen the German (or Ashkenazic) synagogue with its gold *bimah*, ark and second floor banisters, the French (or Canton) synagogue, with its windows overlooking a small canal, so that the New Year ritual of *tashlich*—casting bread onto the waters—could be accomplished from inside the synagogue; and the Levantine synagogue, with its dark wood and red walls and curtains. The Italian synagogue is on the top floor of a private apartment building, and thus generally closed to the public. The largest of the synagogues, the Spanish synagogue, is the one used by the small Jewish community of Venice during the warm months of spring, summer, and early fall—and so it is closed during the main tourist season. But I am in Venice, this time, by sheer coincidence, during a Jewish holiday. Surely, I thought, I could show up at the Spanish synagogue, not as a tourist, but as someone interested in prayer, and be admitted! Of course I was not the least bit interested in actually praying—I just wanted the chance to see the inside of the synagogue. A clever plan.

And so here I am, listening to the torah reading, surprised that several young boys, obviously in training for their bar mitzvahs, are up on the *bimah* and participating in the service. My wife and I were warned, upon entering, that we could not simply leave after looking around—once inside we would have to stay until the conclusion of the service. How long would that be, I thought? An orthodox service, I supposed, probably lasts longer than the American Conservative services I was used to—but this was only the morning prayer service and they had to break for lunch. My wife and I, before we parted ways at the top of the ornate stone stairs that led to the second-floor sanctuary—she to the left, I to the right—made a whispered agreement to leave and rendezvous outside in an hour.

Chapter Two

As it turned out, neither time nor boredom proved to be problems at all. The service was remarkably quick-paced, with all the members of the congregation continually aware of what prayer was being said, so that there were no delays while people found their places. I myself was never lost. The basic structure of a Jewish service is the same the world over, I imagine, and I was able to know which prayers to say—if I felt so inclined—at precisely the right time. And there is clearly something to be said for conducting the service entirely in Hebrew, for although I would have been bewildered by any sections of the service spoken in Italian, I was quite comfortable with the Hebrew. I began to realize the power of a universal common language for a people and a culture.

But there was much more than the language and the beauty of the building that moved me, that made this a major experience of enlightenment, an experience I find difficult to put into words, and yet an experience that I believe has relevance for my work in coming to understand the philosophical meanings of the Holocaust, of culture, place, and nature, and of domination and autonomy. Here also, I begin to consider and to understand a concept that I will call authenticity. And I discovered, after rejoining my wife, that she had felt the same kind of ineffable insight, that something in this specific Jewish service in this specific place, had a special emotional meaning, a special power, for her as well as for me. As with the Warsaw cemetery, the Majdanek death camp, and the sight of Anne Frank's tree behind her house in Amsterdam, the direct experience of this place compelled me to think anew about the connections of history to nature and to human evil and destruction.

Perhaps the crucial moment occurred near the end of the service, when the children of the congregation gathered near their parents for a formal blessing. As this moment in the service approached, children of all ages appeared out of nowhere—most had obviously been outside or downstairs in the main anteroom, waiting for the prayers to end. They came in and glided quickly through the rows of pews to where their fathers or grandfathers stood. Kisses, hugs and greetings were quietly exchanged, and then each family group collected itself under one *tallit*, the one prayer shawl of the grandfather or father. I noticed that even the young girls—all below the age of puberty—had come over to the male side of the synagogue. Huddled together each family group received the blessings of the children, of the generations. And I stood alone, in

Thoughts on the Holocaust

some very important sense an outsider, and thought about the survival of the Jewish people.

Here, I thought, are the real survivors of the Holocaust. Whether or not these particular Jewish individuals had lost family members during the war, they were clearly the remnants and direct descendants of a flourishing pre-war European Jewish community, a community that was directly attacked and imperiled by the Nazi plans of the final solution. Italian Jews, of course, were spared the worst effects of the Holocaust. Under Mussolini's reign, Jews were not deported to the death camps in German-occupied Eastern Europe. Only after Mussolini's fall in 1943 and the German occupation of northern and central Italy, did deportations begin—the first from Rome in October 1943. At the outbreak of the war in September 1939, Italy had a Jewish population of approximately 57,000. Between eight and fifteen thousand Italian Jews perished, that is, between fourteen and twenty-six percent—the best survival rate of any European nation that fell under German control. However, my knowledge of the specific Jewish losses from Venice is sketchy, at best. The historian Lucy Dawidowicz claims that there were approximately two thousand Jews in Venice at the start of the war,[1] and an exhibit in the Hebrew Museum in Venice states that only about one hundred Jews returned after the war. But it is not clear that the Nazis killed all the missing nineteen hundred—and my research has not yet turned up any more concrete figures.

But the precise numbers are not that important, at least not in this case, as an explanation of my experience of this surviving Jewish community. Regardless of the precise calculations of death and survival, the fact remains that these are families of Jews who managed to avoid the Nazi death machine. Watching the blessing of the children, I can feel a deep and palpable authenticity here that is lacking in American synagogues when we recite blessings for the six million. Without denying the fact that many American Jews themselves are direct descendants of survivors of the concentration camps or members of families that fled Europe just before the war—and thus that many American Jews lost close relatives in the Nazi genocide—there is still a basic difference between the American Jewish experience of the Holocaust and the experience of European Jews. No matter the depth of our anguish, sadness, and guilt, American Jews never stood on the precipice, we never looked directly at the face of evil and death confronting us. The Holocaust, as horrible as it was, happened "over there" across a wide ocean in a foreign land.

Sitting in the Spanish synagogue of Venice, a building whose foundation dates to a time before the founding of the first American colonies, I realize that this experience of authenticity can be attributed, to a great extent, to an accident of geography—the source of the experience comes down to the physical location of a place. I am here, in a building, in a city, in a country and continent, with a long and remarkable Jewish presence and history. I am here in a place that as little as seventy years ago was meant (by the Nazis and their followers) to be *judenrein*, purified of Jews, forevermore. And yet it is a place that is not free of Jews; on the contrary, it is a place that exudes the authenticity of the Jewish experiences of history, spirituality, family, community, and survival.

The feelings that I experience sitting in the Spanish synagogue of Venice would seem, at first, to be quite different than the emotional resonances and meaning I ascribed to the other Holocaust sites I have visited: the Warsaw cemetery, the Majdanek death camp, and Anne Frank's tree. After all, there is no experience of the natural world, the nonhuman environment, in the synagogue. Thus, there should be no question about the possibility of the healing power of nature, the restoration of ecological systems, or especially, the liberation and autonomy of natural processes. Yet what is common to both sets of experiences is a strong sense of place, let us say, a rootedness to a particular location, and a connection to a specific historical record. It is that strong sense of place and history that I seem to experience as "authenticity."

II.

It is necessary, then, to examine the connections among place, history, and this concept of authenticity. Since my starting point is the field of environmental philosophy, I begin with an analysis of the meaning of place derived from the realm of environmental studies. Philosophers concerned with the ethics of environmental policy have examined and analyzed the value associated with particular places: bioregions, ecosystems, or wilderness areas, for example. Perhaps the foundational text in a discussion of the connection between place and environmental policy is Kirkpatrick Sale's *Dwellers in the Land: The Bioregional Vision*,[2] but the deep ecology movement has also emphasized the importance of place

Thoughts on the Holocaust

and community with the land. Bron Taylor, for example, claims that "bioregionalism provide[s] the deep ecology *movement* with the *social philosophy* that any comprehensive philosophy must develop."[3] Almost without exception these investigations have concentrated on the important connection between place and nature, on the idea of humanity living and acting in a landscape, a part of the natural world.[4] The possibility that the meaning, value, and ethical significance of place can be found, instead, in a particular human culture and a specific human narrative history is not an idea that dominates discussion in environmental philosophy. But my experience in the Spanish synagogue in Venice suggests the concept of place cannot be separated from an appreciation of human culture and history. Our experience of place can be both an experience of nature and/or an experience of human history and culture.

This is not an earth-shaking conclusion, of course, and it surely comes as no surprise to those who study cultural geography and/or anthropology. But it may be—and I emphasize the word "may" here—a new way to approach philosophical problems in the various fields of environmental ethics, philosophy of technology, and Holocaust studies. What can we learn if we focus our attention on the experience, the existence, the presence, and the meaning of place as it intersects with ideas about environmental destruction, with ideas about the Holocaust, and with ideas about the nature and meaning of artifacts? More specifically, how does the concept of place affect the idea and experience of authenticity? How does our experience of authenticity, in turn, alter our ideas of value and meaning in human activities—especially the activities of creation and destruction—regarding both the natural world and the practice of genocide?

Recent analyses of the concept of place focus on a critique of an "objective" definition or evaluation. Place is considered to be primarily an idea that arises through a subjective experience or a human interaction with a physical location. Thus Daniel Berthold-Bond notes that the very vagueness of the ideas of "region" or "place" is a positive aspect of the philosophy of bioregionalism,

> because it invites us to question purely objective, geographically literal definitions of place ... [B]ioregionalism subverts the mathematical, topographic, literalistic definition of place as objective geographic location ... and develops a new geography of place as experiential, subjective, and meaning-laden.[5]

Chapter Two

Jeff Malpas makes essentially the same point by contrasting the historical origins and development of the ideas of "space" and "place." Space is mathematical extension, while place is an open-boundaried location. The scientific objectivism that has dominated the western world since the scientific revolution has emphasized the objective and quantifiable notion of space. Place is relegated to a secondary and derivative status, for it incorporates affective and emotional experiences into the evaluation of the physical world.[6]

The subjectivity of place means that our ideas of places arise in an active relationship with the physical environment in which we live. The very knowledge that we have about the world is conditioned by our geographical location. "Perception adjusts to the demands of the environment," argues Christopher Preston (following Yi-Fu Tuan), citing the example of the different perceptual abilities of the Inuit in the Arctic or the Kalahari bushmen. Visual clues to observation are useless in the vast whiteness of the Arctic, and so the Inuit hunter relies on sound and smell. The Kalahari bushman, in contrast, uses minute perceptual clues to identify the leaves of plants that have edible roots. Thus knowledge and belief differ in formal structure according to physical location. Preston writes: "Disciplines such as cultural ecology and cultural geography ... have long suggested a dialogical relationship between landscape on the one hand and cultural organization and belief on the other."[7] According to Christoph Rehmann-Sutter, place is a "texture of relations."[8] Culture—knowledge, belief, value, norms—all arise out of a relationship between humans and their environments.

But culture implies a history. In criticizing the idea that places can be described in an objective scientific way, Rehmann-Sutter observes that "places seem to be locations whose nature is that they appear when we address ourselves to them in a distinct way."[9] We look at a physical space but understand it in a new way as a "place" with special relational meaning to us. So the beach that I once viewed as a playground with waves (when I was a child) I now understand to be a distinct (and special) dynamic natural system involving the interactions of humans, animals, plants, and physical processes such as wind, water, and sand. The physical beach is essentially the same spatial entity, but the beach is now a place for me, indeed, a home, a source of rootedness—it is Fire Island. As Rehmann-Sutter writes: "Places are observed through their history.

Thoughts on the Holocaust

They have a meaning which is established by their natural inhabitants who lived and live in them."[10] But they also have a center—a set of relationships that are more intense and more important than others—and the preservation of this set of relationships becomes a normative goal. The people who recognize a physical location as a place that is the source of special emotional, cultural, and historical relations will seek to preserve the existence of the place. Normative claims derive from the historical significance of places. It is from and within history that places are imbued with value.

The crucial lesson to be learned in the analysis and understanding of place is this connection between history and normative claims. And again, I do not think that this point should come as a surprise. Holmes Rolston, perhaps the most important environmental philosopher in North America, has long argued that narratives about natural entities, species, and ecosystems—that is, natural histories—are the primary source of human obligations to the natural world. "We ought to live in storied residence on landscapes," he writes, playing on the double meaning of "story" as both a narrative and a level of a building, a residence, a home. "Each locality, each ecosystem, is unique." Each has its own story to tell us, and ultimately it is only that story—that natural history—that can justify a policy of preservation. "There is no logic with which to defend the existence of elephants or lotus flowers, squids or lemurs; but each enriches Earth's story. That alone is enough to justify their existence."[11]

Janna Thompson has used a similar argument for the justification of the preservation of natural environments. Thompson examines the idea of "cultural heritage" in order to create an analogical argument between the preservation of cultural artifacts—such as historical buildings—and the preservation of natural entities, systems, and locales. History again is the crucial link. Thompson begins with the educational importance of the preservation of historical sites: "the educational effect of heritage depends on the belief that things can have value because of their association with past people." But the argument goes beyond educational value: "Experiencing these things [i.e., historical sites] enables us to remember and honor people of the past and thus to connect ourselves to our history."[12] Thompson emphasizes the importance of historical narrative, the story told about ourselves, as being both the legacy of those things of value that have been passed on to us, but also those things of value that we pass on to the future.[13] The value and significance of an entity "is determined by our historical narrative—the role that something has played in our history and in the lives of people of the past."[14] And the foundation of this value and significance in history is the result of the multi-generational character of human activities: projects, goals, values, and aspirations "transcend" the scope of individual human lives.[15] Natural locations as well as human built environments can meet these criteria for historical and cultural preservation. What is crucial is not whether or not the entity to be preserved is natural or not, but what role it plays in our understanding of our history, whether or not it connects us to the values of the past and future.

This investigation into the meaning of place shows, then, the importance of human interaction with the environment. A place is more than a mere physical location because it has meaning for a set of human actors. The meaning it has as a place arises out of and is constituted by human interactions with the physical environment, and the significance that these interactions attain as part of an historical narrative. This is not to say that all evaluations of place are based on anthropocentric—human-centered—value. The story of a natural species or the history of an ecosystem or bioregion may gain its value because of its special qualities that lie outside of direct human benefit. But we humans—who, after all, are the ones doing the evaluation and instituting the policies of preservation or development—must believe that the story, the narrative, or the history of these nonhuman realms and natural entities is interesting, that they and their story are worth preserving. Natural physical spaces exist outside of human evaluation, but places in nature are defined by the human interest in a specific locale—to repeat Rehmann-Sutter, "when we address ourselves to them in a distinct way." That distinct way, I believe, is as part of an historical narrative that involves both human institutions and natural processes.

Here is where my problems begin: once I consider the importance of human culture and history in the experience of place, I need to re-examine the difference between natural places and places of human development, a difference that is the focal point of my arguments concerning the ethics of restoration ecology. It is in the analysis of restoration ecology that the ideas of environmental ethics and philosophy of technology are brought together in a particularly pointed way, for we begin to examine the nature and value of artifacts that humans create as they intervene in the processes of the natural world. As we saw above in Chapter One, restoration ecology is a relatively new field of environmental management, in which one studies and attempts the rebuilding of degraded ecosystems. For those who believe in the tenets of environmentalism, this appears to be a worthwhile enterprise, repairing the damage that human activities have caused natural systems. But an ethical and political problem is raised by the very fact of restoration ecology: as our scientific and technological ability improves so that we can more perfectly re-create ecosystems, there will be less and less reason to preserve the natural systems that we still have. Why not strip-mine a mountain for its coal and then rebuild and restore the mountain to its original state later?[16]

I have argued that the restoration of a natural system cannot restore the natural value of the original system. The authentic value of the original system has been lost. In this context, I define authenticity as the combination of the originating causal process of an entity and the historical continuity of the entity throughout time.[17] Consider an example from human art and culture: when we refer to a painting as an authentic Vermeer, we mean that it was actually painted by Vermeer (sometime between 1645 and 1675) and that it has not been altered by any subsequent human being in the last 350 odd years.[18] Similarly, an authentic natural area would be one that was formed by means of natural processes (not human technology) and that has not been significantly altered by non-natural processes throughout its history. When humans restore a natural system, we create an artifact that resembles nature—and may, indeed, perfectly capture the essential functional elements of the original—but it differs fundamentally and ontologically from a natural entity or system. Natural entities and systems are not the result of human design; artifacts always are. The restored mountain ecosystem, for example, might look exactly like a natural mountain ecosystem and may function in a similar way—but it will have a different causal history. The natural mountain ecosystem possesses a kind of authenticity—through its causal origins and history—that can never be captured by a human-made artifactual mountain. Thus I call restoration ecology "the big lie." In its praxis, restoration ecology is claiming that human activity can create the authenticity of nature, but this is false, for only natural processes can create natural authenticity.[19]

Chapter Two

But here in the Spanish synagogue of Venice we have a situation in which it is the specific human place that creates the authenticity of the experience. In my analysis of environmental policy I tend to stress the authenticity of natural processes, but in my analysis of the survival experience of the Jewish people I discover that it is the artifactual creations of human culture—the Hebrew prayers, the ritual of the family blessing, the aesthetic presence of the synagogue itself—and the precise historical facts about human events in this particular place that are the source of authenticity and value. In thinking about the natural environment, it is the idea of a genuine unmodified nature that determines my ideas, but in thinking about the Holocaust, it is human culture and history that dominates. Is this a real conflict in my views?

I think not. The key point is that certain types of modification and change will create an historical narrative or continuity that is not authentic. The modification can be within a system of natural processes or in a series of human and cultural events. The issue is the authenticity of the change. Within a natural environment or system, the addition of human technology or human-induced change is a clear disruption of the natural historical process, and thus the result is a lack of authenticity. That is why an ecological restoration of, say, a strip-mined mountain has less value than an unmodified natural system. Within human institutions and events, the determination of authentic change may be more problematic. Humans alter their physical space all the time, and why should we consider some of these modifications to be detrimental to the value of authenticity? I can give no general answer to this problem, but I can point to specific cases where it would be clear that authenticity has been lost. Assume, for example, that after the morning service in the Spanish synagogue in Venice, I talk to some of the congregants as we exchange holiday greetings. In expressing my admiration for the building and its connection to the history of the Jews in Venice, I am told that aerial bombing during the Second World War had destroyed the original synagogue, and that the beautiful building I had been sitting in was constructed in the 1950s. In this scenario, I think that something of my "authentic experience" of an historical Jewish place would be lost. And although my imagined example of a destroyed Spanish synagogue is a mere thought experiment, an almost exactly similar case does exist in the city of Warsaw. There the Old Town square was completely de-

Thoughts on the Holocaust

stroyed by the German army in the spring of 1945, and what exists now is a fully re-constructed replica or facsimile of the old medieval buildings. When I learned of this on my visit to Warsaw, I realized that the point of visiting the "historical" sites of the Old Town could not be to see the original buildings, but at best, to understand the history of how the original buildings were destroyed and re-created. One is unable to experience the authenticity of Old Town in Warsaw. The authenticity of an historical place rests, in some sense, on the continuity of the physical artifacts that comprise the area. Although the Spanish synagogue of Venice, I am sure, has been modified over the last 435 years—it has been cleaned, repaired, and had modern plumbing and electricity added—it is still the same building that was built in the sixteenth century.

So let us return to those places described in Chapter One where the experiences of nature and the Holocaust intermingle, the death camp at Majdanek and the Jewish cemetery in Warsaw. There I was impressed with the power of nature to re-assert itself in places of human evil and mass murder. As I stood in the Warsaw cemetery, filled with lush undergrowth and trees covering over the massive Jewish graves, or as I gazed at the grass growing in the parade ground at Majdanek, I considered the power of nature to heal the wounds of human destruction. Although these were places in which human cruelty would forever be associated, places that gain their primary significance as sites of human mass murder, it was clear that in time nature could—if not prevented by humanity—obliterate all traces of the human evil. As I argued above, nature acts on its own, expressing its freedom and autonomy through the transformation of sites of human history. Although we humans cannot accept these developments of natural history to be a complete healing of the human wounds of destruction, because the causal history of the sites will always include the evil human acts that occurred there, we can still appreciate the power of a liberated and autonomous nature acting without regard to human interests and concerns. Although we need to believe, as did Anne Frank, in a nature independent of human evil and destruction, it is perhaps impossible for us to think of even this independent and autonomous nature completely divorced from human history. Even Anne's tree only gains significance because it is the tree that Anne Frank, a famous victim of the Holocaust, perceived and thought about during her period of hiding. So the idea of a natural place developing through history and indeed interconnecting itself with human history cannot be eliminated from our understanding of the events of the Holocaust. Both nature and human civilizations have histories, and there is no conflict in asserting that these differing kinds of histories are central to the process of evaluation to determine authenticity. In short, there are connections among nature, place, and history that can be understood in terms of authenticity.

III.

Let me introduce technology into this analysis. Although I noted in the beginning of Chapter One that science and technology were essential elements of the Enlightenment project to make the world more comfortable for human life, so far the role of technology in the domination and control of nature and humanity has remained in the background. Yet technology is a major connective tissue among nature, place, and history. Obviously, human science and technology are used in the modification and management of natural processes, from the medical treatment of disease organisms to the agricultural cultivation of the land for food. A project of ecological restoration depends on a proper scientific understanding of the ecological system that is being restored, and it employs various technological means to achieve the desired results: controlled burns, the movement of earth, or the planting of vegetation, as examples. Technology is also present in the Holocaust cases we have been examining: the death camp at Majdanek incorporated the notorious gas chambers and crematoria of the Third Reich, and even the establishment of the Warsaw ghetto was accomplished by the simple technological process of building a wall as a permanent enclosure. Indeed, it is fair to say that the principal way that humans act in their attempt to control the world—nature or other humans—is through technology. Human history, in many ways, is a history of technology. Thus a detailed examination is required of the meaning and normative value of technology, for we need to understand how technology is used as an instrument of oppression and domination.

Yet once we begin to examine the value of technology we encounter an immediate objection. Perhaps the oldest commonplace about the nature of technology is that technological artifacts are inherently neutral or value-free. Humans create technological objects for a specific range of purposes, but the actual use of the technology is subject to the intentions

Thoughts on the Holocaust

of the user. These intentions, of course, may be good or they may be evil, but whether good or evil, the technology itself is neutral: the technology has no purpose, no value of its own, except insofar as it meets the needs and requirements of the agent who employs the technology. (As the now infamous slogan of the National Rifle Association proclaims, "Guns don't kill people; people kill people.") The gun is a neutral technological artifact, to be used for good or evil purposes. When the gun is used to threaten and kill someone who is being robbed on the street, it is being used in an evil way; when the gun is used to threaten and kill a would-be murderer entering my bedroom window, it is being used in a good way. The physical properties and action of the gun are similar in both cases but the intentions of the user are different. Thus the gun, the technological artifact, is value-free, morally neutral. Only in its use does value emerge.

The idea that technology is neutral is pervasive in contemporary society, and indeed, throughout the mental landscape of practitioners of technological design and operation. As a philosophy professor at a technological university, I know firsthand that almost all of my students (future engineers, architects, business managers, and computer scientists) and most of my technology and science colleagues on the faculty subscribe to the view that the creation of technological objects is a value-free enterprise. Many popular authors and academics who write on the subject of the philosophy of technology and engineering, such as Samuel Florman, Melvin Kranzberg, Emmanuel Mesthene, and Joseph Pitt all support the traditional neutrality of technology.[20] Although their arguments differ in many respects, all claim that the main problem in guiding technological development is the evaluation of human intentions and goals. Evil technology is not the problem; evil human beings and misguided social policies are.

Nevertheless, the vitality of this tradition is somewhat surprising, given that a broad and powerful critique of the idea of technology's neutrality has been around for at least a half of a century in the writings of Lewis Mumford, Jacques Ellul, and their followers.[21] The popular view that science is value-free has also come under attack in this time period. Indeed, one way to characterize the post-modern age in which we live is by acknowledging as a basic idea that all human creations—both ideas and physical artifacts—are the products of a particular culture and his-

Chapter Two

tory, and that they are endowed by the creative process with the specific values and purposes of the culture or sub-culture (race, class, gender) that created them. No human creation is morally neutral or value-free because each is the product of a particular culture and world-view at a specific time in history.

Here I do not present any new abstract argument or theoretical critique of the idea that technology is value-free. One purpose of this book is to examine the ways in which technology is involved in the domination of nature and humanity. But if technology were deemed to be neutral in value—value-free—then there would seem to be little point in a critical analysis of its role in the process of domination. Yet by looking at the actual history of the Holocaust we can examine the validity of the idea that technology is value-free. The history of the design and operation of the Nazi death camps provides us with a perfect example of a technology that is not neutral. The physical objects that constituted the structure of the camps, as well as the organizational system that operated the camps, were human creations, designed with a set of specific purposes in mind. These purposes were evil, as is well known—but more importantly, the evil of the death camps was designed into the technological artifacts themselves. The death camps were not, as the commonplace idea might suggest, morally neutral artifacts that were simply used in an evil way. The death camps were not value-free, and as human-created technological systems they thus stand as a powerful counter-example to the idea that technological artifacts are morally neutral. The technology of the death camps was the physical embodiment of genocide. Thus we will be able to see the role technology plays in the oppression and domination of the human, and natural, world.

One of the clearest arguments against the moral neutrality of technological artifacts is presented by the political theorist Langdon Winner, who argues that the neutrality idea is based on an illegitimate emphasis on the separation of the creation of artifacts from their use.[22] The commonplace—or traditional—view is that the "making" or creation of the technological object is morally neutral, and that the value of the artifact only arises when the artifact is used. Winner does not deny that there is some truth to the separation of "making" and "use," but his analysis—which I can merely summarize here—shows that the traditional view overemphasizes the separation and mistakenly uses it as a complete explanation of the issues surrounding the neutrality of technological artifacts. Thus we can say that although there is a distinction between the making of technological artifacts and their use, this distinction should not be the only point we consider when we examine the question of the moral value of a particular technology. The value of a technology does not rest solely in its use, but its creation is also imbued with particular values that partially determine the overall moral worth of the technology.

Winner's argument has both theoretical and historical components. The theoretical analysis of the ways in which technological artifacts are actually used shows that the concept of "use" is too narrow, for it rests on a restricted sense of technology as merely the "tool" of human activity. Technologies are not simply used by human beings as tools—rather, they profoundly and fundamentally re-structure and re-shape human life and society. Winner appropriates a phrase from the philosopher Ludwig Wittgenstein (who was talking about language) and calls technologies "forms of life," in that they become embedded in human activity. (He also cites the argument of Karl Marx that the mode of production determines the form of life in that society.)[23] Winner argues that "as technologies are being built and put to use, significant alterations in patterns of human activity and human institutions are already taking place … [so that] … New worlds are being made." By "new worlds" Winner does not mean new planets, of course, but new patterns and organizations of human life: "The construction of a technical system that involves human beings as operating parts brings a reconstruction of social roles and relationships."[24] Winner concludes, then, that technologies are much more than neutral tools to be used for good or evil purposes. "As they become woven into the texture of everyday existence, the devices, techniques, and systems we adopt shed their tool-like qualities to become part of our very humanity."[25] Technologies become the form and structure of human life, on both the individual and social level.

Winner uses this theoretical analysis of the essence of technology as it relates to human life and activity to argue for a more comprehensive method of evaluating the good (or evil) of technological artifacts and systems, what we can call the "value" of technology. Since technologies restructure human life, it is imperative that we examine insofar as we are able the potential directions of this restructuring before it takes place.

The evaluation of technology has to be more thorough than the traditional (and inadequate) evaluation of "impacts and side effects." What are the fundamental changes to human life and human social institutions resulting from a particular technology?[26] The value of a technology is more than its consequences in human activity—there is more to the value of a microwave oven than its quick re-heating of last night's leftover pasta. The microwave oven changes the way that humans prepare food and organize dinner, a family and social event. Winner asks us to focus on the ways in which technologies alter human life, so that we may see that they are not neutral tools but value-laden systems that create new forms of human reality. Technological value resides in the re-shaping of human life and human institutions.

Winner justifies his analysis by means of several historical examples. One kind of case is that in which a technological device or system is used to solve a political or social problem.[27] Winner tells the story of the master planner for New York State in the early and mid-twentieth century, Robert Moses, who designed the overpasses on the state parkways so that they would not have enough clearance for buses and large commercial traffic. Moses specifically wanted to prevent the poor and working class families of urban New York City from traveling to, and using, the state parks. Since buses could not drive on the parkways, only people with access to private automobiles could gain entrance to the parks. The technology in this case—the design and creation of the highway overpasses—was not politically or morally neutral. The technology was intended and designed to enforce a social hierarchy.

A second type of case is one in which a technology is inherently political in a specific way: it cannot exist or function without a particular moral and political system. The clearest example Winner presents is the technology of the railroad—an example he borrows from Friedrich Engels.[28] Although we might think of the components of the railway system—locomotives, passenger cars, tracks, switches, depots, baggage handling procedures—as independent technological artifacts, the truth is that these all comprise a system in which each element requires the others to function in any effective way. And more importantly, the system must be organized with a specific power structure. Railroads must be run by means of an authoritarian power hierarchy to insure that there are regular schedules. Regular schedules prevent accidents; they are also

Thoughts on the Holocaust

necessary for passengers. One cannot imagine a railroad system that ran on democratic or non-authoritarian principles, under which, for example, each train crew decided when they were going to leave the terminal and when they were going to arrive at the next stop. The technological system requires an authoritarian power system, much more strongly than say, the highway system with private automobiles—although even in the highway system some authoritarian rules are built into the system, such as driving on the proper side of the road.

Thus, for Winner, the important conclusion: "to choose [certain kinds of technology] is to choose unalterably a particular form of political life."[29] In other words, more broadly, technologies determine the forms of human life, and thus the values that humans live by. As Winner's examples demonstrate, this determination of value occurs in two directions. One can begin with a deeply held value, such as bigotry against the poor immigrant class, and then choose and design a specific technology so as to impose this specific value on society, as in the example of the parkway overpasses; or one can choose a specific technology and thereby alter human life in accordance with the operation of the technology, because the technology requires specific human behaviors and human organization, as in the example of the railway system. In both kinds of cases, technology is not morally and politically neutral, for its very design and functionality requires or imposes moral and political values. In Winner's short hand phrase, "artifacts have politics."[30] Technology structures human life.

IV.

The design and operation of the Nazi death camps is another compelling example that technological artifacts and systems are not morally and politically neutral. Given the uncontroversial evidence of the horrors and evil of the death camps and the policy of "the final solution" to the Jewish problem,[31] it might seem that the value-laden character of Nazi technology and science is so obvious as not to merit a serious or prolonged discussion. Yet the words and arguments of Albert Speer—architect, master builder, and armaments minister of the Third Reich—belie this initial and obvious conclusion. In a telling passage from his memoirs, Speer reflects on the connections between the Third Reich and the political and moral values of anti-Semitism and the mass

Chapter Two

killing of the Jewish people. He begins by stating that he gave "no serious thought" to Hitler's hatred for the Jews. Why not? His answer is clear and direct: "I felt myself to be Hitler's architect. Political events did not concern me."[32] Here then is an explanation based on the political and moral neutrality of the technological enterprise of architecture. As the mere architect, involved with the design and creation of buildings, Speer cannot be concerned with the political and moral meaning of the things he produces for the master he serves.

Yet Speer's analysis of his role in the Nazis' technological system of mass murder is much more subtle and multi-layered than this initial statement implies. Bearing in mind that everything that Speer wrote after his release from Spandau prison is to some extent self-serving, it is interesting to see that he does not insist on the moral and political neutrality of technology as the justification of his innocence; rather, he seems to accuse himself of guilt because he did not overcome this technological and professional neutrality. His self-accusation has two parts. First, he reports on his initial attempts to rationalize his participation in the genocidal policies of the Third Reich by making a general claim about totalitarian systems that possess a massive organizational structure and technological capability:

> [I]n Hitler's system, as in every totalitarian regime, when a man's position rises, his isolation increases and he is therefore more sheltered from harsh reality; that with the application of technology to the process of murder the number of murderers is reduced and therefore the possibility of ignorance grows; that the craze for secrecy built into the system creates degrees of awareness, so it is easy to escape observing inhuman cruelties.[33]

But second, Speer refuses to use this professional and organizational isolation as an excuse for his guilt. "It is true," he writes,

> I was isolated. It is also true that the habit of thinking within the limits of my own field provided me, both as architect and as Armaments Minister, with many opportunities for evasion. It is true that I did not know what was really beginning on November 9, 1938 [the so-called *Kristallnacht* pogrom], and what ended in Auschwitz and Maidanek. But in the final analysis I myself determined the degree of my isolation, the extremity of my evasions, and the extent of my ignorance.[34]

Speer here clearly accepts the traditional view of the neutrality of technological artifacts and systems, in that he acknowledges that one could simply think and act "within the limits of [one's] own field"—i.e., as an

Thoughts on the Holocaust

architect or an engineer solely concerned with technical problems and tasks—and thus one could ignore the broader social, moral, and political realities of the technological project that is the focus of one's professional attention. Speer's new wrinkle to this old argument is that it cannot provide the exculpation that he seeks—for in his view he had an obligation to overcome the isolation of his professional and technical focus.

In this chapter I am not concerned with debates over Speer's guilt or innocence. I have introduced Speer's analysis of the rationale for his actions as Hitler's professional architect and armaments minister merely to show the pervasiveness of the traditional view that technological systems and artifacts are morally and politically neutral. (I will, however, return to the question of the moral guilt of Speer and other technological professionals in Chapter Six.) My claim here is that Speer and other supporters of the traditional view are wrong about the neutrality of the technological project of genocide that lay at the heart of the Nazi regime. The technological and organizational system that was created by the Third Reich to exterminate the Jewish population of Europe—the ghettos, the slave labor camps, the railroad transportation systems for the prisoners, the gas chambers and crematoria for the extermination camps—all this was not a neutral artifact or tool to be used for good or for evil; it was inherently evil in its politics and moral values in precisely the ways demonstrated by Winner's theoretical analysis of the origin and nature of human technologies. Thus we will be able to see the role of technology in the domination of nature and humanity.

Let us consider first some examples of technologies within the Nazi system that display values regarding human life. Once one examines the histories of these artifacts, one can see that they embody the values of the Nazi political and social agenda. The design of these technologies is based on a given set of social, political, and moral values that we now judge to be evil. The technological artifacts themselves thus carry within them the values; they are anything but neutral.

The engineers who designed the furnaces for the crematoria in the death camps had specific purposes in mind as they developed their designs. The ovens for the crematoria in the prisoner-of-war and concentration camps were originally meant to handle the bodies of those inmates who died through "natural" causes—malnutrition, disease, overwork, etc. But as the policy of the final solution of European Jewry became actualized, the machinery of the death camps was modified to reflect overt goals of mass killing. As Jean-Claude Pressac and Robert Jan van Pelt tell the story, the original design for a crematorium furnace for Dachau in 1937 was for a massive single-muffle (i.e., a single chamber or retort) furnace decorated with a marble neo-Grecian pediment. But Kurt Prüfer, the engineer for Topf and Sons (the firm that eventually built the furnaces for Auschwitz and Birkenau) realized that crematoria at concentration camps did not need the aesthetic displays of a marble Grecian pediment.[35] Moreover, furnaces with multiple chambers for incinerating the corpses would be more efficient; in the camps, there would be no need to preserve the integrity of the ashes, as there would be in a private, commercial funeral-crematorium establishment. In the death camps, there would be no grieving family to collect the remains. "Prüfer convinced Bischoff [Karl Bischoff, the SS officer in charge of the construction of Birkenau] to create the necessary incineration capacity at the POW facility by grouping three incinerating crucibles in a single furnace."[36] Thus, throughout the history of the design of the crematoria at Auschwitz-Birkenau, we find furnaces with ever-increasing numbers of muffles or chambers for the incineration of corpses, from two to three to a double-furnace with four chambers each.[37] The driving motivation for designing these kinds of crematoria was the anticipated mass killing of Jews and other undesirables. As Pressac and van Pelt note, "the men in the WVHA [the Economic and Administrative Office of the SS] had begun to associate the 'final solution to the Jewish problem' with the capacity of the new crematorium—or crematoria."[38] Of the meeting in August 1942 involving the chief engineers and camp commander where the final plans for the various crematoria at Birkenau were discussed just prior to the commencement of construction, Pressac and van Pelt write: "It was clear to all participants in this meeting that crematoria IV and V were to be involved in mass murder."[39]

The conversion of the original crematorium building at Auschwitz—generally known as crematorium I—to its new use as a combination gas chamber and crematorium is also instructive. Sometime in the fall of 1941 the first experiments with the use of Zyklon-B gas as a mass-killing agent were conducted in the basement of Block 11 at Auschwitz, when approximately 850 prisoners (600 Soviet POWs and 250 Poles) were gassed. For reasons based on efficiency and secrecy, it was determined that the basement killing chamber was not appropriate. The basement required up to two days to air out after the use of the poison gas; moreover Block 11 was in the southern corner of the camp, so other prisoners could see the prisoners entering the building and the removal of the corpses. Thus, the morgue room in the crematorium I building—located behind the SS hospital and across a road, away from the main compound of prisoner barracks—was converted to a gas chamber.[40] Doors to connecting rooms were sealed, and openings were drilled into the ceiling so that the Zyklon-B tablets could be poured into the room. A ventilation system was added or modified so that the gas could be extracted.[41] Using the morgue room in the crematorium building as a gas chamber also had the advantage of a continuous source of heat from the furnace, for Zyklon-B vaporizes at 27 degrees centigrade, and the basement of Block 11 would be too cold for an efficient production of gas.[42] We can assume safely that the consideration of the furnace heat to create the appropriate air temperature for the gas chamber rooms was also an issue in the design of the crematoria at Birkenau, where all four crematoria-gas chamber complexes were built as one building with a central furnace room. In sum, the design of the gas chambers and crematoria were meant to maximize the efficiency and secrecy of the killing operations. The victims were brought to one building alive and were gassed and incinerated out of sight from the rest of the camp personnel and prisoners.

Efficiency in the incineration of corpses can also be seen in the layout of the Birkenau camp, especially if we compare it to the site plan of Majdanek, a camp constructed earlier. As we saw in Chapter One, Majdanek was situated on the outskirts of the major Polish city of Lublin, and thus was the only extermination camp in sight of an urban population. The location itself is thus a peculiar feature of the camp design. Moreover, the schematic site plan of the Majdanek camp shows that the crematorium was on the opposite end of the camp from the gas chambers.[43] Cremation pyres in an open field were near the gas chambers, so this means that most of the corpses from the gassings were burnt in an open field, on the side of the camp closest to the city walls. In addition, there was no railway line into the main camp, so prisoners had to walk along the road or across an open field to the entrance, where selections were made in a small open area next to the gas chambers.

Chapter Two

Birkenau was planned better. The railway line entered straight into the heart of the camp. Prisoners disembarked halfway between the entrance and the four crematoria-gas chamber complexes. Those selected to work in the camp were immediately brought inside, while those selected for killing were marched up the road to the gas chambers. The four new crematoria that began operating in the spring of 1943 all contained the gas chamber and furnaces for the incineration of the corpses in one building. Although the provisional gas chambers—the red and white farmhouses—did not have crematoria and thus made do with cremation pyres, these were some distance from the prisoner barracks and within a wooded area.[44] At Birkenau, then, the site plan, the architectural design of the gas chamber buildings, and the furnaces themselves all were planned with the mass killings of the prisoner population in mind. The increased efficiency of the killing operation was due to the intentional planning of the technologies to be used.

Architectural design in Nazi Germany is, indeed, a fertile area for examining the relationship of political and moral values to the creation of technological artifacts. Paul Jaskot's detailed history of the building program of the SS demonstrates that "specific forms" of architectural design served the political interests of the SS and also that formal design decisions were "functionally instrumentalized for other, seemingly non-artistic goals." Among these were the "oppressive policies of the Nazi state"[45] as well as the promotion of a public identity for the SS and a policy of accumulating power in the German state.[46] I do not have the time, space, or inclination to review Jaskot's many examples, but we can note the design and construction of the SS barracks at Nuremberg, the design of the entrance to Buchenwald, the castle fortress of Wewelsburg, and the stone-quarry labor camps of Flossenbürg and Mauthausen.[47] The design of both the Nuremberg barracks and the entrance to Buchenwald were meant to convey a "unified and monumental presence" and indeed the barracks entrance was based on a model of a Roman triumphal arch, thus achieving the "function of projecting a monumental role for the SS."[48] The castle at Wewelsburg was one of several medieval sites acquired by the SS administration. The SS saw itself as an "elitist racial institution" and thus its members were to consider themselves as an outgrowth of "older traditions of aristocratic service" dating back to Henry I of Saxony. The castle at Wewelsburg was meant to "resurrect" this Germanic past,

Thoughts on the Holocaust

and it was part of the overall plan by Himmler and the SS to establish pre-modern Germanic agricultural settlements in the newly occupied lands in the east[49]—a plan we first encountered in Chapter One.

Finally, we can see in the designs of the labor camps at Flossenbürg and Mauthausen a use of stone building materials to create not only a monumental presence reflecting the "permanence of the SS" but also a "means ... of destroying as many prisoners as possible."[50] Unlike the vast majority of labor and death camps, which were built with an economy of materials and with few aesthetic considerations,[51] the camps at Flossenbürg and Mauthausen were designed so that public SS buildings—including the watchtowers—were constructed of stone, in such a way to be more than merely functional. At Flossenbürg "great care was taken in cutting the stone" for the door frame arches, the windows and the corners of the towers.[52] Designed and built during the early years of the war, these monumental stone structures represented the "height of SS optimism and confidence in a German military victory"[53] and served the dual function—through the brutal working conditions of the quarries—of hastening the destruction of the enemies of the German state.[54] Thus we see again that the design and creation of specific technological artifacts—in this case, architectural structures—is not neutral but embodies specific social, political, and moral values.

V.

The architectural examples of the previous section serve as a bridge to a more comprehensive view of the ways in which technological artifacts and systems incorporate the values of the social and political order. Rather than see the design and creation of a specific artifact as embodying a specific purpose (and hence value) as in the case of the multi-chamber furnaces in concentration camp crematoria, it is possible to see that technologies and systems reify or operationalize a particular world-view, as in the restoration of medieval castle fortresses (and the building of stone fortress-like concentration camps) as symbols of the historical continuity and permanence of SS ideology.

Thus historian Michael Allen has argued that no distinction should be made within the regime of the Third Reich between the "rational pragmatism" of "normal" engineers or "technocrats," who supposedly made technological choices based on efficiency alone, and the "Nazi fanaticism" of the SS administration, who made decisions based on the pursuit of social and political ends.[55] In an historical analysis that lends crucial support to the philosophical argument presented here, Allen argues that a distinction between the pure technocrat and the political ideologue can only be maintained if we consider "machines" or "modern management" techniques in isolation from their "social and cultural history." For Allen, choices about the management of SS operations and the employment of specific technologies "inherently involve choices among different visions of 'community' and 'society,'" and this is because "artifacts commonly assumed to be 'value neutral' never appeared as such to the SS."[56] The artifacts and technologies were always embedded in a culture or world-view that was the driving force of SS activity.

What was the ideology of the SS? Allen notes five central ideals:[57] (1) the remaking of Europe into a New Order, in both the physical sense of creating new communities (and eliminating unwanted inferior peoples) and in the spiritual sense of inculcating a new value system; (2) the Führer principle, under which leadership and national unity became the primary organizational rule; (3) the right-wing socialism of National Socialism, or what Allen terms "productivism," the idea that community good was to outweigh individual profit and that the real purpose of business was to "make Germans and Germanness, … an indelible national harmony" of workers and managers;[58] (4) a belief in the potential of modern production techniques, such as Fordism and Taylorism; and (5) a belief in Aryan racial supremacy.

Among the administrators of SS businesses, these five ideals produced a "managerial consensus" based on ideology. Allen provides many examples of SS business operations—including, of course, the administration of the slave labor and extermination camps—being driven by ideological considerations rather than the normal considerations of business and commerce. One noteworthy example is the selection of the Spengler brick-making machine for the German Earth and Stone Works, a brick-making machine that used a new "dry press" technology. The dry press technique was more expensive than the older "wet" presses and required much more skilled labor to operate efficiently. As Allen notes, "it is hard to imagine anything less suitable for low-skilled, slave labor than the Spengler system." Then why was it chosen? Because of the SS fascination with modern technological systems; the Spengler brick-making machines were chosen "for their symbolic character as icons of modernism."[59] The machines symbolized that the SS was at the vanguard of a New Order.

The selection of modern sewing machine technology in another SS industry (the Textile and Leather Utilization Company or TexLed) was also driven by ideological considerations—but here at least, the machines worked well with a population of slave laborers. The new machines were not labor-saving but "output-multiplying"—with the new machines, each worker could produce more. These new machines were particularly suited for unskilled labor—they required only a short training time and thus could be used in areas with high turnover rates of the labor force. In a climate of brutality as in the concentration camps, these modern new textile machines actually were efficient in the production of textiles, because workers could be forced to produce more.[60] There were also issues of gender involved here. Although it was part of the ideological goal of "productivism" to produce a highly skilled German worker, the textile industry was considered "woman's work" and so modern machines that used an unskilled work force were deemed acceptable.[61]

A final example concerns the use of technological choices to accelerate the SS goal of the extermination of inferior peoples. The construction of the underground building sites for the German rocket industry at Dora-Mittelbau was in part driven by the idea of "extermination through work." Here labor-intensive but output multiplying machines were used to make the underground tunnels—tools such as hand-held drills, hammers, and pneumatic shovels. These were more practical than power shovels, which were capital intensive. It was much more efficient to work to death the slave laborers who used the simple hand tools to bore the tunnels and clear the stone rubble. The system had the additional benefit of serving as a threat to the skilled laborers who worked in the rocket production; they knew if they did not work efficiently they would be assigned to the underground tunneling crews.[62]

In all of these cases, the motivation for technological and business decisions was the ideology of the SS—the social, political, and moral values that the SS wished to generate in the New Order that they envisioned. Allen's thesis is that the operation of the SS business corporations is the best "counterexample" to the idea that business management is "inherently pragmatic" and free of political and social ideology. Sometimes, the ideological considerations produced an efficient business enterprise—as in the TexLed case. Sometimes the ideological considerations did not—as with the Spengler brick-making machine. For Allen then, "the modern management of technological systems 'works' only when both managerial consensus (based on ideology) and sound knowledge of the complex material realities of production can be brought together in a coherent system."[63]

VI.

Jacques Ellul, the visionary critic of our contemporary technological society wrote that "technique has become the new and specific milieu in which man is required to exist, one which has supplanted the old milieu, namely, that of nature."[64] Ellul meant in part that we are embedded in a technological world, that all of our decisions reflect the requirements of the technologies that we use, and that our technologies structure our world. Or, as Winner writes, technology has become a "form-of-life." In such a world, a world pervaded by technology, it is impossible to maintain the illusion of the old traditional idea that our technological artifacts and systems are merely neutral tools for the pursuit of human goals. As Winner demonstrated, there are two distinct ways in which technology is not value-neutral, two ways in which "artifacts have politics." First, technological objects can be designed to embody a political or moral purpose; second, technological objects and systems often require specific forms of social and political organization, reflective of the broader culture they inhabit. The design, creation, and use of technology in Nazi Germany exhibit the lack of value-neutrality in precisely these ways.

As we saw, Speer, the master builder and administrator of the Third Reich, used the idea of the neutrality of technology to explain his participation in the horrors of the regime. In speaking of the technical experts who worked with him to increase the productive capacity of Germany during the war, he wrote:

> Basically, I exploited the phenomenon of the technician's often blind devotion to his task. Because of what seems to be the moral neutrality of technology, these people were without any scruples about their activities. The more technical the world imposed on us by the war, the more dangerous was this indifference of the technician to the direct consequences of his anonymous activities.[65]

Thoughts on the Holocaust

Although Speer appears to be somewhat ambivalent about the truth of the general claim of technological value neutrality ("what seems to be the moral neutrality of technology"), it is clear that he failed to see that the technological choices of the Third Reich were inherently connected to the practical and ideological goals of Nazism. The tools of the Nazi regime, from the furnaces of the crematoria, to the stone towers of Flossenbürg, to the slave-laborers of the underground rocket production tunnels at Dora-Mittelbau, were all imbued with the values of the Nazi project. Individual technological objects were designed with a specific purpose meant to further the goals of Nazism, such as the elimination of the Jewish people. Technological systems were created, organized, and operated from within a specific Nazi world-view, and thus they actualized the values of Nazi culture to create a New World Order.

In sum, the history of Nazi technology provides a convincing counter-example to the idea that technology is value-neutral. Technology can thus be seen as an instrument of oppression and domination, not as a mere tool to be used by human beings, but as the actual embodiment of evil values. The case of the Nazi use and development of technology can be generalized into the historical narratives of other genocides and other forms of domination and oppression. Because we live in a technological world, a technical milieu (as Ellul calls it), we should be able to discover that the forces of technological development structure and drive policies of oppression, domination, and genocide in general.

And so here we can return to a consideration of the human relationship with the natural world. The technologies used to control, modify, or manipulate the natural processes of the environment are also the embodiment of specific values. It remains to be seen whether or not these values are evil, but at first glance, they appear to be anthropocentric, based on human interests and concerns. The values embedded in the technology of environmental management or ecological restoration thus challenge what I have called the authenticity of nature, for they produce systems that lack a continuous causal history of natural processes, and they instantiate human goals. In creating a new technological history of nature, a set of natural processes dependent on human science and technology, humanity imposes its will on nature. The result is the oppression and domination of nature. My conclusion is that the concept of authenticity, the reality of technological development, and the unfolding of history all contribute to our understanding of domination—and this is true whether we speak of the domination of humanity or the domination of nature.

So where does this leave me? Well, still sitting in the Spanish synagogue of Venice contemplating the meaning of Jewish survival. But why, the question might be posed, is there a *Spanish* synagogue in Venice, Italy? The answer lies once again in an understanding of history—Jewish history—specifically in the expulsion of the Jews from Spain at the end of the fifteenth century. Within a few generations the exiled Jews of Spain were making their home in the Venetian ghetto and building the most elaborate and largest of the many synagogues there. To understand this place—its beauty, its meaning, its significance—one must know this history. The knowledge of history, it appears to me, is thus the key notion that binds together the idea and experience of authenticity in the fields of environmental philosophy and the philosophical study of the Holocaust. Authenticity is based fundamentally on history.

In the study of nature and the evaluation of the ethics of environmental policy, authenticity is bound to the history of a continuous natural process of a place, ecosystem, or bioregion. That is why, as I noted above, the technical restoration of a degraded ecosystem is not authentic—it disrupts the causal flow of natural processes and a natural history and replaces it with a human-made artifact. In the study of human activity, authenticity is bound to the history and culture of a place, and to the human experiences of the place with its particular history. These two kinds of histories can, of course, intermingle. Nevertheless, when we study the philosophical meaning of the natural environment or the Holocaust, our ideas acquire authenticity to the extent that we experience the places relevant to these histories. To understand the value of human activity in the natural environment, we need to experience the authenticity of natural places. We need to understand and appreciate the natural histories of specific environments, the history of human actions that alter, disrupt, destroy, and preserve nature. (Here I think the writing of Holmes Rolston is the model for how this should be done from a philosophical perspective.) To understand the Holocaust, we need to experience the authenticity of Jewish life and death, we need to experience the particular history of the places in which the Jewish people lived and died. This means that we must visit the death camp sites, not as witnesses, not as pilgrims, but as seekers of knowledge. At the death camps, we must wander through the museum exhibits that present the details of the Nazi death machine, study the engineering and architectural plans of the camps, and stand on the same ground as the victims. We must also study Jewish life—the historical buildings, such as the Venetian synagogues, the remnants of Jewish ghettos, and the still vibrant Jewish communities in places that were once destined to be free of Jews forever. Only then will we have a full understanding of what was destroyed and what survives.

Let me briefly consider an objection to this position. Andrew Light argues that I am conflating two distinct notions of "authenticity"—what he calls a "psychological" account of authenticity as opposed to an "ontological" account.[66] In my analysis of the authenticity of natural areas as compared to ecological restored areas, Light (correctly) claims that I am using the ontological sense of authenticity. An authentic natural area has specific properties, causally determined through history. But in my description of my experience of authenticity during the religious services in the Spanish synagogue in Venice, I am using a psychological sense of authenticity. The experience of authenticity is not connected to any particular objective property of the synagogue or the religious service. There is no authentic entity, merely a psychological state that I experience as I view the blessings of the children in this particular place and time. For Light, then, there is no reason to compare these two forms of authenticity, and no conclusion about the value of authentic natural processes or authentic cultural practices can be gleaned from such a comparison.[67] In addition, and more importantly for Light, the use of the psychological authenticity of experience forces the question, "whose experience counts?" Perhaps a different observer of the religious services feels no sense of authenticity.

Light's objection clearly requires me to emphasize what he calls the ontological account of authenticity and value. My memoir of my experience in the Spanish synagogue is meant to reveal the actual authenticity and value of this particular place in the process of Jewish history and culture, not merely my subjective experience of the place. On my view, since authenticity is tied to history—actual historical reality—the experience of authenticity must be connected to an ontological reality, even if the details of this reality remain inexpressible. It becomes a moral obligation to preserve these authentic places, whether they are natural areas or artifacts of human culture, and whether or not we can pinpoint the precise physical qualities that produce the authenticity. Preserving the authentic ontological reality is a necessary condition for preserving the possibility of authentic experience. This is another reason why the historical existence of Anne Frank's tree is so important—it signified the existence of an independent nature free of human domination, guiding us to an authentic experience of human liberation and freedom from domination. The experience of authenticity is based on the authenticity of history, and both are central to our understanding of domination.

When I travel to foreign countries, I make a point of visiting Jewish sites—the Venetian ghetto is not my only tourist destination. Thus I once found myself in Spain, in the section of Barcelona known as the Call. Until the Jews were expelled in 1424, this was the once-thriving Jewish section of Barcelona. During its heyday in the twelfth, thirteenth, and fourteenth centuries it was the financial and intellectual center of Catalonia, boasting the only real university of the country, the Jewish College. Barcelona was also the home of Moses ben Nahman (known as Nahmanides), one of the most famous of Jewish philosophers. In the year 1263, in what is now known as the Disputation of Barcelona, Nahmanides defended the Talmud against a prominent Christian theologian (Pablo Christiani) who claimed that the midrash supported a belief in Jesus as the messiah. By the early 1400s, the synagogues were closed and the building stones removed to build new palaces for the ruling class of the city. All that remains today, on a single street, the Carrer de Marlet, is a nineteenth-century house bearing a stone in its outer wall that dates from the year 1314. There is a Hebrew inscription, which translated states: "Holy Foundation of Rabbi Samuel Hassardi, whose life is never-ending." This stone is the only remaining physical evidence of the Jewish presence in Barcelona from the medieval period.[68] But the physical evidence is not as important as our memory, our understanding of the history of the Jewish expulsion from Spain, and our experience of the surviving descendants of those Spanish Jews who built the synagogue in Venice. Standing at this spot, one experiences a Jewish place. The life of Rabbi Hassardi is never-ending because he lives now in our memory and our knowledge of history. He lives in the morning service on Simchas Torah in the Spanish synagogue of Venice. He is with us, as are the victims of the Nazi Holocaust, under the prayer shawl, receiving the blessing of families and children.

Thoughts on the Holocaust

Notes to Chapter Two

1. Dawidowicz 1975, 368–71.
2. Sale 1985.
3. Taylor 2000, 273; see also Devall 1988.
4. Two exceptions that emphasize the need to consider the urban environment as well as the natural landscape are Andrew Light (Light 1998, 179–84) and Roger J. H. King (King 2000, 115–31).
5. Berthold-Bond 2000, 7.
6. Malpas 1998, 31–32.
7. Preston 1999, 213; citing Tuan 1974.
8. Rehmann-Sutter 1998, 175.
9. Rehmann-Sutter 1998, 172.
10. Ibid.
11. Rolston 1998, 286–88; see also Rolston 1988: 328–54.
12. Thompson 2000, 247.
13. Thompson 2000: 255.
14. Thompson 2000: 257.
15. Thompson 2000: 249.
16. A more detailed discussion of the normative and ontological issues arising in the practice of ecological restoration will be found below in Chapter Three. Note here that this objection to restoration—that it would make it harder to argue for preservation—is applicable only to nonhuman natural systems. No one would seriously argue that it is permissible to mutilate humans because it is now easier (through plastic surgery) to "restore" their features to a pre-mutilated state. This dis-analogy supports my claim in Chapter Three that there is a difference between nonhuman nature and human culture.
17. Katz 2000, 40–41.
18. What precisely we mean by "altered" in the case of human artifacts—and nonhuman nature—will depend on context. The Vermeer painting may have been cleaned and repaired, but the canvas has not been re-painted. The direct manipulation of the painting surface will always provoke controversy; witness the public debate over the "restoration" of the Sistine Chapel. For a fuller discussion of the spectrum of activities that are involved in restoration, see Chapter Three.
19. Again, for a more detailed discussion of these issues, see Chapter Three below.
20. See Florman 1975, Kranzberg 1991, Mesthene 1983, and Pitt 2000.
21. Mumford 1963; Ellul 1964. One major disciple, Langdon Winner (Winner 1986), will be discussed below.
22. Winner 1986.
23. Winner 1986, 14.
24. Winner 1986, 11.
25. Winner 1986, 12.
26. Winner 1986, 17.
27. Winner 1986, 22–25.
28. Winner 1986, 30–31.
29. Winner 1986, 29.

Chapter Two

30. Winner 1986, esp. 19–39.
31. I am deliberately ignoring the arguments of the Holocaust deniers. For a comprehensive treatment, see van Pelt 2002.
32. Speer 1970, 112.
33. Speer 1970, 112–13.
34. Speer 1970, 113.
35. Pressac and van Pelt 1994, 185–86.
36. Pressac and van Pelt 1994, 199.
37. Pressac and van Pelt 1994, 218.
38. Pressac and van Pelt 1994, 216.
39. Pressac and van Pelt 1994, 219.
40. United States Holocaust Memorial Museum 1996, 95.
41. Piper 1994, 158–59; and Pressac and van Pelt 1994, 209.
42. Pressac and van Pelt 1994, 209.
43. United States Holocaust Memorial Museum 1996, 101.
44. United States Holocaust Memorial Museum 1996, 96.
45. Jaskot 2000, 115.
46. Jaskot 2000, 116.
47. Jaskot 2000, 117–39.
48. Jaskot 2000, 120–21.
49. Jaskot 2000, 123–25.
50. Jaskot 2000, 127.
51. Jaskot 2000, 126.
52. Jaskot 2000, 130.
53. Jaskot 2000, 132.
54. Jaskot 2000, 133.
55. Allen 2002.
56. Allen 2002, 64.
57. Allen 2002, 12–16.
58. Allen 2002, 13.
59. Allen 2002, 68.
60. Allen 2002, 75–77.
61. Allen 2002, 75.
62. Allen 2002, 229–30.
63. Allen 2002, 72.
64. Ellul 1983, 86.
65. Speer 1970, 212.
66. Light 2002.
67. Light 2002, 205–06.
68. Time and the human development of urban tourist sites have caught up with my description of the Call, for since my visit in 1993 an underground synagogue used by the Jews of medieval Barcelona has been discovered and restored. The existence of additional Jewish artifacts at this historical site only buttresses my argument, although it may diminish the poetic nature of the importance of the single memorial stone.

~ Chapter Three ~

ECOLOGICAL RESTORATION AND DOMINATION: THE NEED FOR AN INDEPENDENT NATURE

I.

I have seen destructive erosion of the beach many times in my life, and the impact of Hurricane Sandy in the fall of 2012 was not the worst. As I stand on the Fire Island beach and observe what little remains of the protective dunes, I can at least see that none of the houses in my community have been destroyed. It was far different in the spring of 1992, when a series of severe nor-easters wreaked havoc to dozens of homes.[1] Of course, the perception of damage is extremely localized. Hurricane Sandy is considered one of the worst storms in the history of the New York and New Jersey coastal region, and many other communities fared much worse than where I live. My community was lucky; yet the damage is still considerable. Earlier that summer, I had remarked to a friend that the dunes were in better shape than anytime since 1976—high and wide and covered with a thick and stabilizing vegetation. Now the dunes are completely gone, the sand pushed back and distributed in random piles behind the second and third row of houses. Virtually every plant within 150 yards of the shoreline—grass, shrubs, and mature evergreen trees—is dead, killed by the excessive salt spray of the storm or the rising seawater itself.

  The consolation is that the dunes performed their function, or at least the function imagined by the human population of the coastal community: the dunes were sacrificed to the ocean surges, but the houses of the residents remain. Now the process of re-building the dunes begins, and I am faced—in my own personal life—with a variation of the process of ecological restoration. In the past, the dunes have been re-built in a variety of ways: by scraping the excess sand of the flat beach and pushing it onto the small remaining dunes, then planting vegetation to anchor the sand; by constructing snow fencing to catch the wind-blown sand;

Chapter Three

and by massive sand mining projects that suck sand off the ocean floor a mile off shore and pump it onto the beach. All of these procedures can re-build the dunes as an artifactual product of human technology. None of them, of course, restore the natural contours of the beach. They are, instead, instances of the human attempt to manipulate and control part of the destructive forces of nature.

In Chapter One, I argued that a respect for an autonomous natural world required us to intervene in natural processes, when necessary, in the most minimal way possible. Using a theoretical thought experiment about the re-building of dunes on Fire Island, I concluded that sand replacement was better than building permanent structures such as sea walls and stone jetties, and it was also better than doing nothing, "letting nature be," because Fire Island was a hybrid environment that molded together a human infrastructure with a once natural beach ecosystem. Now, because of Hurricane Sandy, I am confronted with a decidedly non-theoretical version of this problem. As a homeowner in this beach side community, I have a strong incentive to advocate a policy of beach restoration; after all, I want my home and property to be protected. Yet I still maintain that the most minimal intervention for the repair of the dunes is the most desirable policy alternative. I do not want to see permanent sea walls built as a bulwark against the encroaching ocean, for such a massive technological project would change forever the character of the beach community. My real life experience reinforces my view that limits must be placed on human projects of domination.

In the first two chapters of this book I have argued that nature and humanity can have connected histories of domination. I have also argued that technology can embody specific normative values as it literally re-makes the world. In the previous chapter, the discussion focused on the ways in which technology entered human history, with the example of the Nazi death camps, to create a realm of mass murder. Here, I will focus exclusively on the natural world, and the practice of ecological restoration. In my view, ecological restoration is nothing more than an expression of the human domination of nature—even in the case of the re-building of the dunes of Fire Island. By examining this particular form of domination, I will show the significance of Anne Frank's tree as a symbol of an independent nature that can be used to resist and to limit the forces of domination.

Ecological Restoration and Domination

II.

In previous chapters, I have raised some of the philosophical issues of ecological restoration, but here I give a detailed account of the critical arguments and objections. The argument begins with a critique of ecological restoration that I proposed more than twenty years ago, following the work of Robert Elliot. In addition to the idea that the policy of restoration was an example of the human domination of nature, I claimed, more pragmatically, that a belief in the validity of restoration would subvert, and render meaningless, the environmentalist goals of the protection and preservation of natural systems and entities. I remain committed to these basic ideas, despite the appearance of numerous critiques of my original arguments.[2] In this chapter I present a discussion of the arguments against ecological restoration and the objections raised against my position. I have two purposes in mind: (1) to defend my views against my critics, and (2) to demonstrate that the debate over restoration reveals fundamental ideas about the meaning of nature, ideas that are necessary for the existence of any substantive environmentalism. It is this second purpose that connects most with the central theme of this book: the resistance to domination as exemplified by the horse chestnut tree in Amsterdam.

First, we need to review the basic criticisms of ecological restoration. The process of ecological restoration became a philosophical issue with the publication of Robert Elliot's essay "Faking Nature" in 1982.[3] Elliot presented an argument against a hypothetical position that he called "the restoration thesis," the idea that a damaged or degraded natural environment could be restored to its prior status with no significant loss of value. Elliot was concerned that the acceptance of the restoration thesis would lead to an increase in environmental policy decisions that had a negative impact on ecological and natural systems, because it would provide the developers of land with arguments that could be used against conservationists and preservationists. If the natural area or ecosystem could be restored after it has been used—for mining, logging, or agriculture, for example—then why not use the land, reap the economic and social benefits, and then return the area to a prior natural state?

Elliot's seminal argument used as its basis an analogy with art forgery to introduce the robust normative elements of the restoration thesis. A perfect art forgery—if possible—would still lack the value of

Chapter Three

the original artwork because of its genesis and history. Part of what gives an artwork its value is the process by which it came to be. A painting may look exactly like a Rembrandt, but if it were not actually painted by Rembrandt it would have less value.[4] But it is not just the forgery or fakery that is determinative of the value: deception is not the central issue. Elliot also presents a case where a person admires a sculpture only to discover that it has been made from a human bone, and indeed that the human being was murdered precisely so that the bone could be used for the sculpture. The value of the artwork now radically changes.[5] This shows the importance of the causal genesis and history of the artwork for the determination of its value.

For Elliot, the connection to the preservation of undisturbed natural areas or wilderness was clear. What many people value in undeveloped nature is its natural history separate from human causation and activity. In an area that has been modified by human action there is a different causal history. Thus, even a perfect ecological restoration lacks the value of the original natural system it is re-creating, for the restoration was the product of human action.[6] It does not have an origin in strictly natural processes unmodified by humans; it lacks an historical continuity with an unmodified natural system. Elliot concludes that the restoration thesis is unsupportable, and thus it cannot be used to justify the development (and subsequent restoration) of natural ecosystems and areas. The restored area will have less value than the original system.

Over the last three decades this basic normative critique of ecological restoration has grown more complex. In part this is because the conversation between restorationists and their philosophical critics (and defenders!) has shown that the process of ecological restoration is itself complex, with a multitude of forms and purposes. The original case of sand mining followed by a restoration of the dune system that inspired Elliot to question the "restoration thesis" can be seen to be a limiting case at one extreme of the entire array of policies that can be called ecological restoration. The mere clean-up of trash from a meadow or stream can also be considered to be a restoration, perhaps as a limiting case at the other end of the spectrum. Between these two extremes is a wide variety of restoration activities, such as the elimination of exotic plant and animal species, the removal of dams so as to return stream and river courses to their original states, the replanting of blighted areas, and

Ecological Restoration and Domination

the re-introduction of original species to re-create historical landscapes. Many of the restoration activities within this broad middle of the spectrum basically use the processes of nature itself to bring about desired ends; the human activity in these cases is limited, as much as possible, to the mere elimination of obstacles to natural development or the initial re-introduction of natural processes (such as a controlled burn). Indeed, these kinds of cases—in which the restoration is accomplished by nature working to restore itself, rather than a massive human management of natural processes—are primarily used by ecological restorationists to defend the practice against critics such as Elliot. A natural area restored by natural development will exhibit historical continuity with the original natural system.[7]

So the question arises: does ecological restoration remain a philosophical problem? I believe that it does. Although it is clear that a wide variety of restoration projects exist, they all share a common feature that lies at the heart of the normative issue: the presence of human intentionality and design. This common feature calls into question the idea of the replacement of natural entities as a morally justified human policy of action.

Over the last two decades, I have made a series of arguments regarding the normative problem of ecological restoration based on the presence of human intentionality and design. In part, I have simply continued and expanded Elliot's original criticism of the restoration thesis based on the analogy with artworks. Origin and historical continuity—what we might call authenticity—are the crucial elements in the determination of the value of an artwork. When we examine and evaluate a work of art, we want to know who the artist was (or is), and under what conditions and historical circumstances the work was created. With prehistoric or ancient art, where the specific human artist is unknown, we at least want to know the specific time period and geographical region in which the art was produced. A work of art that appeared similar to a work by a specific artist or from a specific time-period or place of origin that was not actually created by that artist or from time period or place would be valued differently. Moreover, as Elliot's human-bone sculpture example shows, the origin of an artwork also concerns the manner and means by which the work was created. There will be disvalue associated with art created by processes that we deem immoral. Historical continuity is similarly important. We want to know that a work of art has had a continuous existence throughout time without any damage and without any changes. Combining these elements of origin and historical continuity yields the condition of authenticity: the work of art we see today really is the same work of art created by a specific artist (or at a specific time and place) in the past, unmodified by subsequent events.[8]

One way in which I extended Elliot's argument was to consider the authenticity of dynamic works of art, such as ballet, opera, or other dramatic works of performance art, since these are more similar to the dynamic processes of natural systems. Origin, historical continuity, and authenticity are still crucial factors in the evaluation of performance art. Consider the recent controversy in New York theater circles concerning a new revival of the American classic opera *Porgy and Bess*. The creative team of the revival discussed adding a "back-story" for the character of Bess and changing the ending of the opera, as well as other changes to the plot, dialogue, and physical movements of the actors. Although these changes may make the opera more accessible to a general audience, more profitable to the producers, and even more enjoyable, the critical factor is that the new production will lack the authenticity of the original: it is no longer the *Porgy and Bess* created by George and Ira Gershwin and DuBose Heyward. New and different elements have been added and original material has been deleted: origin, historical continuity, and authenticity have all been violated.[9]

Shifting back to the restoration of natural systems, the elements of origin, historical continuity, and authenticity continue to play a decisive role in the determination of value. Here however, we add the new elements of human intentionality and design as relevant to the determination of value. In the case of artworks, problems arise because the original artist or artists are no longer the creators of the work we see, but in the case of the restoration of natural systems, there is no original artist or designer. With restored natural systems the problems with authenticity—the break in historical continuity and the change in the causal origin—come about because we add human intentionality and design. We humans interrupt the natural development of an area and modify it to meet human goals and ideals. We attempt to mold the natural system to meet our needs—needs that may be economic, political, scientific, cultural, or aesthetic. We turn nature into an artifact created for human purposes.

Ecological Restoration and Domination

There is a fundamental ontological difference between artifacts and natural entities; they are different kinds of things. Artifacts are created for a purpose. They are the products of intentionality and design. Indeed, artifacts only exist because they fulfill some purpose. They would not be created and produced unless some goal was envisioned for them. This is true even when we consider certain creations by nonhuman animals—such as beaver dams—to be artifacts. Now the characterization of the products of nonhumans as artifacts may be problematic, for it raises a host of issues concerning reason and purpose in the animal kingdom, but I believe we can bracket off these questions without any serious impact on my arguments concerning human artifacts. It is the existence and meaning of humanly created artifacts that is the issue here, and how these human artifacts differ from natural entities. Ecological restorations, after all, are projects that are conceived by human beings. And it is clearly true that human artifacts are created for a purpose, and that they are the products of human intentionality and human-conceived designs. This is completely unlike the origin of natural entities. Natural entities do not exist because of any process of design or purpose, unless one wants to posit a theological design and purpose. Given the validity of Darwinian science, we can safely reject that alternative conception of the origin of natural entities. But note that even if a theological interpretation of the origin of natural entities were accepted, there would still be a difference—a fundamental ontological difference—between human artifacts and the natural entities created by God. Human artifacts would be the result of human intentionality and design, and that would be completely unlike the intentionality and design of a divine being.

The value of natural entities and artifacts is different because of this ontological difference. Unlike artifacts, a large part of what makes natural entities valuable is their freedom from human control. Nature is mostly that wild other realm separate from human plans and projects. This is the sense in which we can say that nature is autonomous, analogous to a human subject in its ability to develop by means of its own internal logic. It is this autonomous development that is modified when we interfere to control the processes of nature. If this autonomous development is replaced with human intention and design, we have a system with a different origin and a different history: we no longer have an authentic natural system or entity. A natural entity or system modified

Chapter Three

or controlled by human intentionality and design has a different value than a natural entity or system that follows its autonomous development.

Artifacts, on the other hand, are the physical manifestations of human intention and design. They are the physical manifestations of human purpose imposed on the world of nature. The value of artifacts derives from the fulfillment of the purposes for which they were created. This means that a project of ecological restoration is not really the restoration of a natural system; it is the creation of an artifact, an artifactual system. Within this system there will be natural entities—so we may be able to call it a hybrid system—but the system as a whole will be the artifactual product of human intentionality and design, created for a human purpose. Now the purpose of a restoration project may be extremely positive, it may be significant and important. Perhaps we are mitigating the damage caused by pollution, or repairing the damage caused by a natural disaster such as a flood or a hurricane—as in my real life case of the beach and dune restoration on Fire Island. We might be re-creating an historical landscape that has both cultural and ecological importance, or re-developing wetlands that had been destroyed by a housing project. These beneficial purposes would tend to justify policies of ecological restoration. But these activities should not be characterized as the restoration of nature: they are not. These activities are the creation of artifactual systems—or at best, hybrid systems composed of natural entities and artifacts. To call the product of an ecological restoration project the restoration of nature is, as I provocatively proclaimed over twenty years ago, a "big lie."

The issue here is not the possible benefits that can be derived from restoration projects—of which we may all agree—but rather the fundamental meaning of the policy of ecological restoration. If we misunderstand the meaning of restoration, we fail to understand the extent of the human impact on the natural environment. We will fail to see the ever-increasing humanization of the natural world, the limitless expansion of human power to mold and manipulate our entire environment. For restoration, despite its good intentions and its support from environmentalists and environmental scientists, is a continuation of the human project of the domination of the natural world. It is a continuation of the paradigm of human scientific and technological mastery over natural processes. This grand human project to control the natural

Ecological Restoration and Domination

world is an attempt to destroy the autonomy of nature, a chief element of its value as that wild other separate from humanity. The underlying assumption of this scientific and technological project is that humanity can control and direct natural processes to better effect than nature can. This viewpoint changes the goals of environmental policy, replacing the ideals of preservation, conservation, and protection with manipulation, modification, and control. Preservation and protection will lose all substantive content; they will be meaningless terms in a world of the unlimited modification of natural processes, a world in which the human domination of nature will be complete.

III.

Consider some objections. Return briefly to a point touched upon in the above argument: the possibility of positive restorations. Andrew Light has criticized my emphasis on the human domination of natural entities and processes—the subversion of natural autonomous development—by highlighting the difference between benevolent and malicious restoration projects. A benevolent restoration is one "undertaken to remedy a past harm done to nature although not offered as a justification for harming nature."[10] Light argues that benevolent restorations can work to restore the autonomy of nature, by eliminating prior human interference. If we simply remove the obstacles for a natural regeneration of an area or ecosystem, then autonomous natural processes will take over and re-create the area or system. In addition, Light argues, "the relationship between humans and nature imbues restoration with a positive value even if it cannot replicate natural value in its products."[11] Restoration activities, for Light, serve as a bridge between humans and nature by creating for humans the opportunities for positive experiences working with natural entities. What is restored then is "the human connection to nature" or "the part of culture that has historically contained a connection to nature,"[12] or "what could be termed our culture of nature."[13]

Eric Higgs has made a similar argument, although his position is based on a distinction between purely technical restorations and those that are similar to the "focal practices" championed by philosopher of technology Albert Borgmann. According to Higgs, "technological restorations" are those that are mainly concerned with the perfection of technique; they feed into the dominant technological culture and lead to the commodification of nature. So-called "focal restorations" on the other hand are "shaped by engaged relationships between people and ecosystems."[14] Within a focal restoration project, the human actors will deeply value the ecosystem being restored and also honor the social relations that are formed through the restoration practice; if they fail to value nature then the end result will be the commodification of the natural system.[15] The key element here, for Higgs, is the authentic engagement with the natural area under restoration; without authentic engagement we run the risk of a merely "denatured" technological fix.[16]

Both Light and Higgs are thus claiming that restorations can be good based on the kinds of relations that are developed between the human restorers and the natural area under restoration. In a sympathetic consideration of this argument, Ned Hettinger casts doubt on the conclusion. Hettinger claims "restoration's positive vision for the human/nature relationship fails as it rests on a prior destructive relationship with nature." Even more strongly, he writes, "the restoration paradigm suggests that the proper role for humans in nature is first to degrade nature, then to attempt to fix it."[17] Obviously, such a relationship of harm-then-heal is not the intentional goal of restoration practices; but Hettinger seems correct that there is something odd about claiming that the attempt to heal anthropogenic harms to nature somehow represents a positive or authentic experience with natural processes. Surely a better positive experience with nature involves no harm at all; and that is why I argued in Chapter One that the best policy humans can have with nature is to "leave it alone." Hettinger, however, like Higgs and Light, wants to find some positive involvement that humans can have with nature. His conclusion is that we must learn "to distinguish between respectful human *use* of nature and human *abuse* of nature"—and only then will we avoid the destructive domination of nature.[18]

My rejection of the argument that restoration can produce a positive experience for humanity is more fundamental. I reject this claim based on the simple idea—developed in section II above—that restored ecosystems or entities are no longer natural beings but rather artifacts. This calls into question the entire notion that humans can have an authentic experience with nature when they are dealing with a restoration project. Working in a garden—feeling the soil in one's fingers, planting the seeds, pulling weeds, and watering the plants—may produce positive

Ecological Restoration and Domination

human experiences, but these are not the experiences of working with natural entities. A garden is not a natural area. Perhaps this is the reason why Higgs spends so much time discussing what he calls "ecocultural restoration"—not the mere replacement of ecological integrity but the building (and re-building) of human community and culture.[19] The idea of restoring nature through human technology and science is simply a non-starter: to justify the process, Higgs and Light need to introduce the benefits that these artifactual re-constructions have for human community and culture. These benefits may be considerable, but they are not the restoration of natural processes. As Higgs notes, "In setting goals for restoration … it is unlikely that human agency will follow history."[20] The historical continuity of a natural area is not an element of the restoration process. We are dealing here with the creation of an artifact.

Indeed, I claim even more radically that working in a garden, rather than teaching us about the authentic experience of natural processes, actually furthers the human worldview of domination. Working in a garden teaches us how to control natural processes; it teaches us how to convert natural processes into an artifactual human project designed to serve human purposes. Gardening is a subset of the discipline of agriculture. The name is telling: we do not call the control of plant life to meet human needs "agrinature." It is a cultural process; it is the creation of an artifactual system. And so with all restoration projects: the underlying lesson is that human science and technology can control natural forces and processes. The underlying lesson is the glory of the human domination of nature.

It is thus a mistake to think that there exist "positive restorations" that somehow create a beneficial experience for humans as they relate to natural entities and processes. Yet if we remove the possibility of positive human experiences as an argument against my criticism of the restoration project, what remains at the heart of this objection is the possibility of a continual autonomous unfolding of natural processes. As I briefly noted above, this is the objection of Richard Sylvan, who argued that not all restorations are artifactual because nature can heal itself.[21] Given enough time, nature can wash out any human influence. Consider a garden plot that has been created by the clearing of a bit of forest. The garden can only be maintained if there is continuous human action, for example, tending to the weeds. If the human maintenance activity ceases, the natural processes of the forest will re-assert themselves, and the area of the garden will become overgrown and wild again. It is true that without continuous human intervention, the future development of environmental and biological systems will be natural, i.e., nature will be autonomous. But we cannot overlook the fact that the progress of the system will be different after the initial human intervention. The resulting system will be different from what would have been the case had no intervention taken place at all. The forest plants that grow over the garden plot will be different from the plants that were removed to create the garden. Following Sylvan, we might not want to call the new forest an artifact, since it is no longer guided by human intentionality and purpose, yet the new system is not equivalent to undisturbed nature. And this garden-forest case is perhaps the most benign example. In a case where we are dealing with the cleansing of pollutants or the construction and then removal of human structures (such as a dam or roadway) it is even more obvious that the resulting area, after the re-emergence of natural processes, will not be equivalent to what might have been.

The defense of ecological restoration based on the power of nature to heal itself is merely a perverse continuation of the idea that humanity can and should dominate nature. We saw one version of this argument put forth in Chapter One, when we considered the healing power of nature at the Holocaust sites in Warsaw and Majdanek. There I suggested that a belief in an all-powerful healing nature is a form of anthropocentrism. The belief that nature is so powerful and beyond human control that it can heal itself no matter what humans do to it is the mirror image of the belief that humanity can control, heal, and restore the natural processes and entities of the world. The belief in an omnipotent nature correcting our mistakes is simply a moral rationalization of the human desire to control natural processes for the furtherance of human ends. This objection to my critical arguments against the restoration project must be rejected. Although nature can develop autonomously after a human intervention into the system, the resulting system will always be different from a natural progression without human interference.

A second objection thus arises: my critique of ecological restoration rests on a dualism between humanity and nature, or more precisely, between culture and nature. This characterization of my position is valid, but I do not believe that the dualism is pernicious or that the acknowl-

Ecological Restoration and Domination

edgement of this dualism undermines my analysis. Indeed, the dualism of artifacts and natural entities is the heart of the argument. Humans have lived for at least the last ten thousand years (since the birth of agriculture) in a cultural world, essentially constructed and controlled by human technology and science. Although we are biological beings, we do not live in nature; we live in an artifactual environment. Although we human beings are the products of an evolutionary process, the things that humans do—what we create, build, imagine—these are all artifactual, with a source outside the realm of naturally occurring entities, processes, and systems. Our artifacts, our culture, our world would not exist if we humans had not intentionally interfered with and molded nature.

The intentional interference and modification of nature is the source of the culture/nature dualism. Nature alone could not produce the world in which we find ourselves. Nature cannot produce a chair. Nature can produce many entities on which I can sit—a rock, a ledge, a fallen tree, a grassy meadow—but without the imposition of human intention and design we will never see nature produce a chair. So it is the presence of human intention and design that separates the world of human construction from the natural world. Nevertheless, this culture/nature or artifact/natural entity dualism is not absolute. The duality exists along a spectrum. Entities can be more or less artifactual and more or less natural. Judgments can be made based on the closeness of the entity to the original natural source, so that a wooden chair is more natural than a plastic one. Or judgments can be made based on the amount or kind of human intentionality or design that goes into the productive process. Thus placing snow fencing on a dune to catch wind-blown sand is more natural than using a bulldozer to create large sand dunes. But both processes (the fence and the bulldozer) are in some sense artifactual; one is just more so than the other.

So an objection to my dualistic perspective is really an objection about the meaning of artifacts and their relation to humanity and nature. Both Yeuk-Sze Lo and Steven Vogel have presented detailed criticisms of my conception of artifacts and the use of this idea in the debate over restoration ecology. Although their arguments are quite different, they share a basic criticism that my position relies too much on the dualism of artifacts and natural entities, and this dualism, in turn, rests on an unclear or even incorrect meaning of the concept of artifacts.[22]

Chapter Three

Lo raises a number of objections arguing for the need of clarification. First is the connection between human purpose and the concept of artifacts in the restoration project. Lo claims that some restoration activities can be undertaken for the purpose of aiding nonhuman species or entities, such as restoring bamboo for the benefit of pandas. This fact undermines my claim that restorations (and the creations of artifacts) are necessarily anthropocentric. "Whether human technology is involved in a nature restoration project is simply irrelevant to whether the purpose behind the project is anthropocentric."[23] Thus artifacts are not necessarily anthropocentric, and the ontological dualism that I use as the basis of my criticism of restoration cannot be sustained.

To answer this objection, note two points. First, let me re-emphasize that the dualism of artifacts and natural entities resides along a spectrum, and that things can be more or less natural or more or less artifactual. So the purpose behind the creation of an artifact—in this case, the restoration of a natural area—can be more or less directed to human or nonhuman interests. It is important to determine the intentional plan of the restoration project. But this just means—and this is the second point—that the intention that guides the restoration can be a direct human interest or an indirect one: it is always some kind of human interest or purpose. Lo's example of the bamboo restoration for the benefit of pandas is telling, for pandas are those cuddly looking charismatic mega-fauna that human beings love to watch, especially on television. It is a complete mischaracterization of the purpose of the restoration project to say that we humans restore the bamboo for the benefit of the pandas; although the pandas benefit from the restoration of the bamboo, the real reason we undergo the restoration is for the benefit of those human beings who like to see pandas. In my view then, all restorations and all artifacts are necessarily created for some human purpose, even if that purpose is indirect. The existence of a spectrum of purpose does not change the essential meaning of artifacts as things tied to human purposes and goals.[24]

Lo also makes an interesting objection regarding the concept of artifact as it applies to the modification and control of human beings. She notes that the dependency of a human being on medical technology does not make the person into an artifact. The fact that John is the product of in vitro fertilization, or that Mary has a pacemaker, does not

Ecological Restoration and Domination

make them into artifacts. "If, as Katz declares, the technological fix of nature merely produces artifacts, don't the medical treatments given to humanity mainly produce artifacts too? ... The absurdity of regarding human patients as a mere artifact appears to be a *reductio* of Katz's assimilation of restored nature to an artifact."[25] Moreover, the reason why we do not regard the human heart patient or technologically fertilized infant as an artifact is that their ontological dependence on human technology and intentionality is only partial; they are essentially biological beings that operate through autonomous natural processes once the technological intervention has done its work. Similarly, then, with restored nonhuman natural systems. After the human intervention into the natural system, after the restoration project, the natural entities that comprise the ecological system will function as autonomous beings, not as artifacts. Lo uses the example of the restored gray wolf in Idaho and Yellowstone Park, some of whom were captured in the Canadian wilderness and some of whom were bred in captivity before being released into their new habitats, as examples of autonomous natural entities that are only partially determined by the intervention of human technology. Once the wolves are released into the wild they will continue to survive only so far as they use their natural biological capacities.[26] Technological intervention by itself does not make a biological being—human or nonhuman—into an artifact, because the technology only partially determines the existence of the entity.

I have also used the case of the re-introduction of the gray wolves, but in a way that subverts Lo's conclusion. I will not repeat the entire argument here.[27] The key point is that we can imagine a range of cases that lead to the re-introduction of gray wolf populations into a healthy and functioning ecosystem: wild Canadian wolves wander into the United States and establish themselves there; captured Canadian wolves are relocated; captured wolves are bred in captivity and introduced into a new habitat; various wolves from zoological parks are selectively bred and their offspring released into the wild; even the cloning of wild wolves that are then released into the wild. Let us assume that all of these cases result in the re-establishment of a healthy wolf population in areas where wolves had been eliminated. All of the cases, in Lo's terms, would demonstrate autonomous natural entities using their natural and biological capacities to survive and flourish, a result that all environmentalists would applaud.

Chapter Three

Nevertheless, the cases are different in their value and meaning. They exist along a spectrum of human technological intervention. The cases have different value because of the amount and type of human intervention. Quite simply, wolves that have been bred in captivity are different than wolves that have always been wild. So there is a sense in which we can say that even autonomous biological entities that have been modified are partially (at least) artifacts; they are clearly different from entities that have not been modified. To return to Lo's examples of humans modified by medical technology, then, we can say that these humans are at least partly artifactual: Mary with her pacemaker is more artifactual than Sally with her original completely biological heart. Surely this is one of the lessons from Donna Haraway's discussion of the emergence of the cyborg human: with the increased development of technology as it modifies human bodies, we are becoming less natural and more artifactual.[28] From tweezing eyebrows to plastic surgery, from pilates to liposuction, we turn our physical selves into artifactual projects. Thus human beings can be considered to be artifacts: it all depends on where the modifications fall on the spectrum.

So it is clear that biological beings can be artifactual: wolves bred or cloned are different than wolves born in the wild. But if one wants to insist that humans are different than nonhuman natural entities and that their modification by medical technologies does not make them artifacts, this does not undermine the claim that nonhuman entities altered by technology are artifacts. To claim that humans are different from nonhuman natural entities is just to re-assert the dualism that is at the heart of my criticism of ecological restoration. I argue that humans and their activities and products are different from the processes of the natural world; that is what dualism means. Thus Lo's attempt to reduce my argument to absurdity by claiming that humans modified by technology are not generally thought of as artifacts is a non-starter. The key point is to recognize that natural entities modified by human technology are artifacts; the status of modified human beings is actually irrelevant to the discussion. The serious dualism that I advocate precludes Lo's use of the human medical modifications as counterexamples. Humanity is different from nature. Ultimately, I believe that this conceptual dualism is necessary for an understanding of what nature means. The dualism is

Ecological Restoration and Domination

embedded in our use of language. I will return to this argument later in this chapter.

Because of the focus on the autonomy of natural entities, even after they have been modified, Lo also claims the restoration of natural areas does not involve the process of design. Because a restoration seeks to re-create a prior state of a system—what she calls a "reference state"—it is merely a copy of prior natural system, and thus not the product of a human design. "The crucial distinction between a copy and a design is that a copy always presupposes a template, whereas a design does not, in that novelty is a necessary aspect of a design."[29] In the case of an ecological restoration, "the template is something naturally evolved rather than designed by humans, therefore the copy of it (a restored natural entity) is not designed by humans either."[30] But if restored natural entities are not the products of human design, then they are not artifacts. Thus Lo can claim that the dualism that lies at the heart of my critique of restoration is without foundation, for restorations should not be treated as different from naturally occurring entities.

But Lo's argument rests on peculiar claims about the essence of design. Why is novelty a necessary condition of design? Although the patent office may require some novelty in a design or invention in order to award a patent, this is not true of the design of almost all artifactual creations. Surely when I plan to create an artifact that is an ordinary object of everyday life—say a bookcase for my study—I have a design in mind. My bookcase will be no different than countless other bookcases, except that it will be comprised of a unique collection of wood, screws, and braces. If I bought the bookcase from a furniture manufacturer with the sales condition that I assemble the bookcase myself, the design will be printed out in a set of instructions that I will meticulously follow. So design does not require novelty: the *nth* iterative copy of any artifact will have a design. Even more importantly, when we turn back to restoration projects, design is clearly evident. When restorationists attempt to make a copy of an original "reference state" they need to have a design, a plan, to accomplish the restoration project. Even if the goal is a copy of a naturally occurring entity or system that was not designed the copy itself must be designed or planned. Are the actions of ecological restorationists simply random and unplanned? No: they work according to a design. Restoration projects are intentionally planned human activities that follow a design in order to reach a goal, the production of a specific entity or system. This product is an artifact.

In sum, I do not find that Lo's objections to my conception of artifacts are compelling. Restoration projects are always guided by human interests and purposes; like artifacts they would not exist if not for a desired human end, even if we need to introduce a spectrum of direct and indirect anthropocentric interests to account for actions that seek to benefit natural entities. Moreover, there is a spectrum of artifactuality when we consider modified natural entities or even modified humans; there is no absurdity in calling a human with a pacemaker (or an artificial heart!) an artifactual being. So too with modified natural entities or systems: even though they are autonomous beings, once they embody human intentionality and design, they become artifactual. And finally, restoration projects are always guided by an intentional design, even if the design is meant to replicate an original state of nature that occurred without a design.

Steven Vogel, from a different perspective, has also raised a series of objections to the dualism of artifacts and nature. Vogel rejects dualism for two basic reasons. First, given the pervasive influence of humans on the natural environment, there is virtually nothing that exists in the world that is separate from human civilization. If what we mean by the word "natural" is that part of the world that exists outside of human interference or modification, there is almost no nature left. The second reason is that humans are entities that have evolved through the biological processes of nature. This means that what humans do is natural, so the creation of artifacts is a natural process. Thus there is no dualism.[31] Everything in the world is, in one sense, artifactual, for it has all been subjected to human interference, and everything that humans do, in a sense is natural, since humans are biologically evolved entities.

This basic argument is supplemented by more specific criticisms. Vogel considers it important—and problematic—that my dualism treats the human species as different from all other biological species and natural entities. "Why, after Darwin, do we treat *this* particular species [i.e., humans], which after all evolved naturally in the same unplanned way as any other, as something outside of nature?"[32] And he cites my use of an argument by Andrew Brennan where I argue that humans act naturally when they act within their biological and evolutionary

Ecological Restoration and Domination

capacities and that they act unnaturally (or artificially) when they act to supplement or modify these natural capacities in order to manipulate or control them.[33] Vogel finds this distinction meaningless or circular: "how could we [i.e., humans] engage in activities that go beyond our biological capacities?" The carbon dioxide we exhale is produced naturally; so too the carbon dioxide produced when we use fossil fuels to power internal combustion engines, because "the building and operation of an engine [is] an expression of humans' natural capacities" unless we have made an arbitrary and stipulative claim that all of technology is unnatural.[34] So humans and human technology are entirely natural.

Vogel then turns his attention to the meaning and nature of artifacts. Although he admits that human creations are artifacts, at least so far as they are intentionally planned and produced,[35] he does not see this fact as a problem for ecological restoration projects. The central idea in Vogel's criticism is that the idea of purpose in the creation of artifacts is problematic, and cannot bear the normative weight needed to reject ecological restoration. The intention and purpose of artifacts is not clear and precise. Many artifacts that are designed for one purpose are used for an altogether different purpose. Often the purpose or the intention of the creator of the artifact is ignored and the artifact is used for some other goal. Thus Vogel argues, "the 'nature' of an artifact is not determined so much by what its builder intended as it is by the way in which it is *used*."[36] This use may be the creation of autonomous systems that lack a specific human purpose. Following the argument of Lo (discussed above), Vogel claims that the intention behind a restoration project might be the creation of a system that would be allowed to develop "without hindrance." The intention here, according to Vogel, is to "transcend intentionality ... humans might intentionally produce a situation that is *out of human control*,"[37] an ecological system guided by its own internal natural processes. Although Vogel agrees that the product of this restoration project would be an artifact, since it is something that has been intentionally planned by humans, it would be a system that developed without regard to human purpose, for once created it would follow its own internal nature.[38]

Here Vogel offers a provocative comparison with the procreation of human children. Building on a comment that I made that not all intentional creations are artifacts—I used the examples of planned pregnancies and human friendships to show that intentionality is a necessary condition of artifacts but not a sufficient one[39]—Vogel emphasizes that one of the primary purposes of having a child is to create an autonomous being with a nature of its own. Even though the child is the result of intentional human activity it does not exist merely for the purposes of the parents; it is its own autonomous being.[40] Vogel then compares the procreation of a human child with the work of Steve Packard, the restorer of the oak-savanna plains of the American mid-west (and an example that I used in my critical essays on restoration.[41]) Packard is quoted as stating that the whole point of restoration is "to set in motion processes we neither fully control nor fully understand, " and Packard himself makes the child/restoration analogy that the goal of our activity (as either parents or restorationists) is to make the created being "more truly itself."[42] Thus Vogel claims that, based on my own admission that not all human creations are artifacts (e.g., children), restoration projects can be grouped into this category of entities that are created in order to follow their own inner direction. Restorations are not artifacts created for the fulfillment of a specific human purpose and thus their normative value can be asserted without reference to anthropocentric interests.

My response is that the comparison of procreated human children and the restored ecological system is at the very least, disingenuous, and more likely, flat out incorrect. Vogel and Packard and other advocates of restoration may talk a good game about their goal of creating a self-directing system outside of human control, but the fact is that Packard (for example) has a very precise idea of what type of ecosystem he is trying to create through his design. Packard is trying to re-create the oak-savanna of the American mid-west before the arrival of European settlers. Similarly, other restoration projects attempt to re-create a specific ecosystem or natural area that existed before anthropogenic changes were introduced. All of this is very unlike what parents do when they "plan" to have a child. With a human child, we really do wish to create a self-directing autonomous subject, and if we have any goals for our progeny, they are quite general and rather vague: may they be healthy, happy, and productive, perhaps. Parents who have more specific goals for their children—such as those who want their child to be a classical pianist, or a major league baseball player, or a physician—and who carefully structure the lives of their children to meet those goals are seen as

Ecological Restoration and Domination

somewhat dysfunctional. These parents who overly plan or design the lives of their children are treating the children as objects—as artifacts—to fulfill their own (i.e., the parents') needs and interests. So if there is any analogy between restoration projects and the procreation of children, it is all on the negative side. Restoration projects appear to be similar to the actions of dysfunctional parents who attempt to over manage and over direct the lives of their children in order to create specifically designed entities (a specific ecosystem or a specifically talented child). The idea that in either case we are designing and creating a self-directing entity free of external control is simply incorrect.[43]

But the comparison does raise the fundamental issue of the normative limits of intervention. We can return briefly to arguments offered by Lo: she argues that intervention in nature is not always destructive, nor is it always disrespectful of the autonomy of nature, and thus it is not always a mode of domination.[44] Humans by necessity have to intervene in nature in order to survive and flourish. Lo argues that we can do this in a constructive way, just as we intervene in the lives of other human beings. Here the parallels with children arise again. How much intervention in a child's life is appropriate? Obviously, part of the task of a parent is to raise a child that will be a mature and autonomous adult. We need to intervene in positive ways even though we limit the freedom of the child. When exactly does good parenting become exploitation or domination? There is no clear answer of course, and this ambiguity is what provides the fuel for the production of countless "how-to" books on parenting and endless advice from other parents, friends, and relations. Is the same problem evident in the intervention in nature?

The value of intervention is clearly the focus and purpose of Vogel's analysis and criticism of my views on restoration. Going beyond the specific objections that Vogel raises about dualism and the meaning of artifacts in my arguments, he suggests a more positive approach to understanding the human moral obligation to act within and through nature. This approach is based on Vogel's notion of wildness, which is not freedom from all human intervention, but rather the existence of unpredictable events beyond our design and control. This is why Vogel believes that restoration projects can "be consistent with … ongoing wildness." Indeed, he claims "to see that the wildness we're after *is there all the time*, throughout the restoration process; it's not something that comes in at the end, not something we *produce*, but rather something that we *use*."⁴⁵ The restorers of natural systems use the wildness because the processes they begin by controlled burns, planting, moving soil, or introducing animal species are all uncontrollable by human technology and science. We can begin the process but then we have to let natural forces and processes take over the future development.

As we have seen above—in arguments by Light, Sylvan, and Lo—this is a claim often repeated by advocates of restoration: wild nature is actually restoring itself. But Vogel takes the point to a new level by arguing that artifacts also contain a degree of wildness, that is, a sense of the unpredictable. "To build an object—*any* object—is to build something that always exceeds one's intentions, that always possesses something of the unpredictable and unknown about it."⁴⁶ A building may crumble; a bridge may collapse; a flowerbed may fail to bloom; an essay may lose its conclusion. There is a wild nature in artifacts, and Vogel attributes this to a "*gap*" between "the intention with which the builders act and the consequences of their acts."⁴⁷ Whatever humans create they use the processes of nature, which cannot be completely controlled, and so all their activities are wild and ultimately unpredictable.

Vogel is quite correct here in the idea of the gap between intention and final product. He is merely putting into a philosophical essay some of the most chilling lines of verse ever written by T. S. Eliot:

> Between the idea
> And the reality
> Between the motion
> And the act
> Falls the Shadow.⁴⁸

Here is a conclusion I contemplate every time I try to write a philosophical essay, prepare a lecture, cook dinner, or hit a tennis ball: there is indeed a wild and uncontrollable gap between the intention and the completed product, between the product and the goal. But what is the normative conclusion that we can derive from this gap? And how does it reflect on the philosophical issue of ecological restoration?

For Vogel the point is that we need to accept the responsibility for our actions in the environment—here meaning a world that is both natural and artifactual co-extensively—and to recognize with humility that much of what happens as a result of our actions is beyond our control.⁴⁹ Accepting our responsibility and humility will lead to a better world because we will understand our human nature, and our limitations to control this world in which we exist. But for Vogel it is important that we understand our human selves and this world as deeply connected. Humans act in and through nature. The normative problem for Vogel is not human intervention in a pure nature, the transformation of nature into an artifact, but rather an evaluation of human activity in the environments and landscapes that we inhabit. For Vogel, we need to act in regard to the environment so that the "activity is engaged in in the right sort of way."⁵⁰

This much is certainly true: we do need to act in the right way. As I have stated in the past: "To be morally justified, all human activity, even that between humans, requires a standard of appropriate intervention. The determination of that standard is the central question of moral philosophy."⁵¹ But where is Vogel's criterion or standard of appropriate action and intervention? Responsibility and Humility—these are criteria that fail to provide concrete moral guidance. What is the right way to act? What constitutes a good intervention, a responsible or humble intervention? The problem for Vogel's argument is that natural entities and artifacts are indistinguishable. Humans and their actions are natural; artifacts contain within them a wild nature. Everything is natural and everything is artifactual. Thus all human activity is simultaneously natural and artificial and we have no way to make distinctions as to what is good and bad for nature or humanity. Vogel does wish to avoid those actions that have made the world "ugly" or that have been "ecologically harmful"⁵² but given the seamlessness of the natural and artifactual worlds, under what criteria or standards do we determine ugliness and beauty or ecological benefit and harm?⁵³ For the fact is that humans can use their technological prowess to make artificial entities immensely beautiful—a polluted lake, devoid of all life, can be crystal clear and aesthetically pleasing. And the concept of "ecological harm" loses all meaning in a world where human technology and science can re-create, restore, and manage natural processes.

Now Vogel has claimed that the proper method for determining appropriate actions regarding the environment would be through a process of democratic consensus. We cannot rely on "nature" to offer us a normative guide, because the natural and the human are co-extensive:

"the human and the putatively 'natural' worlds are inextricably intertwined to a degree that makes it pointless and indeed conceptually incoherent to try to distinguish them, because the relation of humans to the environment is fundamentally active and transformative."[54] Thus "we cannot find a criterion for environmental judgment in *nature*—because our only access to nature is one mediated by practices through which the environment has already been transformed by us."[55] So Vogel claims that we must evaluate these practices by which we transform and come to know the natural world. But how do we evaluate the practices? How do we make a beautiful and sustainable world? This, for Vogel, is "irreducibly a *social* and *political* question"[56] that requires an answer in democratic decision-making.

There are two critical problems, however, with the idea that the proper criterion for human action in the natural environment should be the result of the democratic process. First, as Vogel has framed the issue, the considerations that we humans use to determine the appropriate activity will necessarily be anthropocentric interests. How could there be any other interests, for in Vogel's view of the world there is no nature—it has been entirely transformed by human activity? But second, if we only consider what humans want in the active transformative interaction with the environment, there is no reason to think that political choices will lead to a better, more beautiful, or sustainable world. The social and political consensus could very well be a world that environmentalists find abhorrent; indeed such seems to be the case, if we open our eyes and survey the world around us.

The prospect of ecological restoration projects is a prime example of these problems. For the process of restoration exhibits the technological mastery of the natural world as it creates landscapes pleasing to the human community. The re-building of the sand dunes near my community on Fire Island will be such a project. The goal of the Fire Island project, as with all restoration policy, is not the preservation and protection of nature. Restoration projects thus lead us to the conclusion that the entire world must be conceived of as an artifactual system, the result of human transformations and action. To resist this thoroughgoing humanization of the world, we require a principle or ideal that can stand in opposition to human power and human interest, so that, for example, we can minimize the amount of human intervention and

Ecological Restoration and Domination

control. The principle or the ideal we need is that of a nature that exists independently from human culture. In this book, this ideal is symbolized by Anne Frank's tree, as she wrote: "Nature is the one thing for which there is no substitute!" The conclusion then is that we must preserve the distinctions between humanity and nature, between artifact and natural entity, so that we have a normative principle to check the power and limit the scope of human domination.

IV.

Up to this point, I have defended my original criticism of the project of ecological restoration from several fundamental objections. These objections have been wide-ranging and have included many specific claims and counter examples, but they mostly converge on a distinct theme: that I have over-emphasized the distinction between humanity and nature; that I have misrepresented the meaning of artifacts as distinct from natural entities; and that I have misjudged the normative value of the distinction. As a consequence, they conclude, I have failed to see the value of autonomous nature acting on its own throughout the restoration process. There is some irony in this series of objections, for one of my principal philosophical beliefs is that there exists an autonomous nature, analogous to a human subject, which must be respected and preserved. So my critics have attempted to use my own thesis, in part, against me.

In the preceding section of this chapter, I believe that I have answered these objections. Here I would like to add another argument for the importance of maintaining the distinction between artifact and natural entity, an argument based on the conceptual apparatus we require for understanding the world. In short, this is an argument about the use of language.

The linguistic use of the term *nature* is obviously ambiguous, and countless authors since the time of J. S. Mill (at least) have noted that we use the term in two basic senses: first as all that exists in the universe, and second as all that is nonhuman. It is clearly the latter sense that is important for environmentalism, because nature in the first sense, as all that exists, cannot be destroyed or even harmed. But it is the existence of nature in the sense of all that is nonhuman that is questioned by Vogel, for example, in his double-sided claim that nonhuman nature no longer exists and that all human artifacts possess a degree of wild nature. So it is the existence of nature in this second sense that is the crucial issue.

Recently Helena Siipi has analyzed in more detail the meaning of *natural* and *unnatural* as it relates to this issue as well as to normative problems in medical ethics and biotechnologies. The result of her analysis is a complex taxonomy of the meanings of natural and unnatural used in a variety of contexts. She notes that *natural* can be applied to various kinds of entities: objects, beings, traits, events (including actions), and states of affairs.[57] There are also different reasons why we attribute naturalness or unnaturalness to these kinds of entities: based on history, or the properties, or the relations between entities.[58] Moreover, we determine naturalness or unnaturalness through two conceptual frameworks of modal degree: whether naturalness is conceived as a "continuous gradient or an all-or-nothing affair" and whether naturalness is conceived as all-inclusive.[59] These different categories of understanding naturalness or unnaturalness are combined to yield the various and manifold cases where naturalness is a problem or issue. For example, a history-based reason for considering an entity natural, in that it is totally independent from human activity, if conceived as an all-or-nothing affair, will yield Vogel's position regarding the end of nature: no such entities exist because of the pervasiveness of human transformative activity. Siipi thus concludes: "in practice, it is not useful to adopt naturalness in [this] sense … as an ideal of biological conservation" for naturalness in this sense is "unattainable."[60]

I will not review here all the applications of Siipi's taxonomy, but simply note two consequences that are relevant to the argument I am proposing in this chapter and book. First is the idea that general discussions about the meaning of naturalness or nature are inappropriate, and probably meaningless, because there are a manifold of ways in which we can understand natural and unnatural. When discussing naturalness or unnaturalness we need to discuss the specific form of the term being used.[61] And second, we need to stress the idea that in most cases, natural and unnatural must be understood along a gradient or spectrum. Judgments about natural value must be based on specific concrete cases, which can differ in degree, not abstract and universal categories.

This methodology works to positive effect in Siipi's further analysis of the meaning of *artifact* as this term is applied to the debate over

Ecological Restoration and Domination

ecological restoration policy. Siipi begins with the intentional modification of entities, since this seems to be a necessary condition for an entity to be considered an artifact: "the properties of any artifact have been intentionally modified by a human being or by a group of humans."[62] But especially when considering biotic entities, not all modifications are sufficient to make an entity into an artifact: adding one sunflower to a field does not make the field into an artifact, nor does adding a ski track through a snow-covered forest.[63] What is needed to make a modified entity an artifact is that the intentional action of the human being brings the artifact into existence by causing it to have certain properties.[64] For Siipi, this will distinguish the problematic case of the human infant from a typical case of the manufacture of a chair. It will also eliminate some cases that have been cited as counterexamples to my general argument against restoration projects. For example, a stream polluted by human industrial activity is not an artifact by Siipi's account. Although the pollution is the result of intentional human activity, the stream did not come into existence because of the human modification of natural processes.[65] This analysis forces me and my critics to focus the debate over the artifactuality of ecological restoration on specific restoration projects themselves, not on the general modification of natural entities.

Siipi makes a further distinction between artifacts and side effects. Sawdust or pollution, for example, can be the foreseen consequences of intentional modification and the creation of artifacts, yet they themselves should not be considered to be artifacts, for the purpose of the intentional action was not to create the side effect. Siipi notes an essential element of artifacts that is substantively equivalent to my view: "artifacts are never just expected and foreseen, but always the goals of the activities by which they are produced."[66] An artifact, as I have argued, is always the result of some intentional human purpose; the artifact would not exist without the desired end. Side effects exist because of human activity, but they are not the purpose of the activity; they are not natural entities, but they are not artifacts either. Siipi argues that this analysis of artifacts based on intentionality and purpose means that the important distinction we should consider is between artifacts and non-artifacts, not between artifacts and natural entities. Non-modified entities, whether living or not, fall into the class of non-artifacts. Siipi gives the examples of zebras, dandelions, waterfalls, and boulders. But more importantly, focusing on this distinction can explain why damaged ecosystems are not artifacts: although humans modified the natural state of the polluted stream by intentional activity, the pollution was not intentional; the human purpose was not to create a polluted stream.[67]

In addition to the intentional creation of an entity, Siipi cites the role of function as a second condition in the meaning of artifacts. It is not enough for a new entity to be created by human intentional activity, but the new properties that are caused by the human modification must result in a new function. The combination of the bringing into existence condition and the new function condition are, for Siipi, sufficient to make any entity an artifact.[68] I believe, however, that this combination of conditions is too narrow, for I have different intuitions about several of the examples that Siipi cites, such as a beautiful stone that one uses as a paperweight or genetically modified corn that is more resistant to pests.[69] Siipi considers neither of these cases to be artifacts: in the first case of the stone there is no creation or modification (unless we broaden the idea of modification to extend beyond the physical) and in the second case of the genetically modified corn there is no new function created; the modification "only makes it more suitable for the functions for which it is currently used." For Siipi, only if the modified corn was given a new function—say it was genetically altered so that eating it would reduce cholesterol—would the new corn be an artifact.[70]

Although I have doubts about some of these examples, I think it is clear that the distinctions noted by Siipi help to clarify issues in the analysis of ecological restoration. Indeed, Siipi concludes her analysis of the meaning of artifacts by generally supporting my use of the concept in the description of restoration projects. Her conditions work to justify my claim that most intentional restoration projects involve the creation of artifactual systems. If an industrial developer destroys a forest but then re-plants and re-builds the ruined area to create a new forest, we have an artifact: an intentionally created new entity with a new function. The function is new because the re-created forest has a different function than the ruined area that existed prior to its restoration. It also has a different function than the original forest, since part of the reason why the re-created forest was produced was to atone, in some sense, for the damage to the original forest. Thus the new system is an artifact. But not all intentional modifications of an ecosystem would be

Ecological Restoration and Domination

artifacts, for if the damaged system still retained its original function, then modifications—such as the remediation of pollution—would not be enough to consider the restored entity an artifact.[71] This analysis of the artifactuality of restoration projects supports my claim that we need to analyze restoration by means of a spectrum. Restoration projects may be more or less artifactual because of the kind and amount of new functions that result from the restoration activity.

In sum, Siipi has developed a linguistic analysis of the meanings of *naturalness* and *artifact* that tends to support my critique of ecological restoration. By emphasizing intentionality, purpose, and function as part of the essential meaning of artifacts her analysis places most restoration projects in the realm of artifactual systems. By noting that there are different kinds of natural and artifactual systems her analysis makes explicit the importance of viewing this categorization along a spectrum, or a gradient, of naturalness and artifactuality. Thus her analysis permits me to avoid criticisms of my view that claim that my characterization of restoration projects as artifacts is too broad: I can, for example, claim that restoration projects are artifacts while at the same time permitting the simple remediation of damaged ecosystems. Nonetheless, even in Siipi's narrow view of artifacts (a view with which I do not necessarily agree), most restoration projects will be artifactual because they involve more than remediation—they involve the intentional modification of systems and areas.

The success of Siipi's linguistic analysis as a means for understanding the philosophical issues in restoration policy suggests that we can use arguments about language to address even more fundamental questions in this debate, most notably the problematic status of the dualism between humanity and nature. I claim that the conceptual dualism of humanity and nature is a necessary condition for any meaningful philosophical or policy analysis of the ethics of environmentalism. In making this claim I am following the seminal argument of Kate Soper: "the *a priori* discrimination between humanity and 'nature' is implicit in all discussions of the relations between the two."[72] Soper sees this conceptual distinction historically: "an opposition … between the natural and the human has been axiomatic to Western thought, and remains a presupposition of all its philosophical, scientific, moral, and aesthetic discourse."[73] Whether we take a social constructivist (or anti-realist) view of the meaning of

Chapter Three

"nature" as something that humans create, or we adopt a view that sees humanity as "part" of nature, we assume the background of the conceptual distinction, if only to argue against its existence.[74] The distinction also remains as the foundation of all discourse about environmental policy. According to Soper, "all ecological injunctions"—i.e., whether to pursue nonanthropocentric goods at the cost of sacrificing human interests, to leave nature alone, to develop sustainable policies to conserve natural systems, to safeguard future resources—all these policies are "clearly rooted in the idea of human distinctiveness."[75] There can be no denial that this distinction exists and forms the basis of our thoughts regarding the environment. "What is then at issue in the humanity-nature division is not the positing of the distinction in itself, but the way in which it is to be drawn, and importantly whether it is conceptualized as one of kind or degree."[76]

The dualism of humanity and nature—the conceptual distinction between them—is a question of grammar, the fundamental use and meaning of the terms. Paul Keeling makes a convincing argument for this point in an essay defending the preservation of wilderness. Critics of the wilderness idea, claims Keeling, cite the mistake of positing a human-nature dualism as the central philosophical objection. A belief in the existence of wilderness is based on an "idealization of pristine, untrammeled nature [that] enshrines an untenable human/nature dualism."[77] This is the same objection, it must be noted, that has been lodged against my criticism of ecological restoration, particularly my use of the distinction of artifact from natural entity. Keeling claims that "the objection is a red herring"[78] relying on a poor analysis of the meaning of nature and an avoidance of the real normative issue of the value of wilderness areas.

Keeling begins his argument by a criticism of the strategy of attempting to find an essential meaning to the term *nature*—a criticism leveled at both my views and Vogel's rejection of my views. The attempt to determine one essential meaning of nature (and its supposed opposite, artifact) leads to either questionable ontological problems if one follows my argument or to Vogel's "unhelpful generalization that all artifacts are natural."[79] Instead of attempting to find one essential meaning we should consider the performative aspect of speech about nature and artifacts, so that we see that what is involved here is a "certain kind of rule-guided

Ecological Restoration and Domination

practice" about the *use* of the words *nature* and *artifact* rather than an analysis of meaning. This Wittgensteinian approach recognizes the obvious "multi-faceted and complex usage of the term 'nature'" but unlike the abstract criticism of dualism, it places the use of the terms nature and artifact in context. A person claiming to love nature "is ordinarily not specifying a special fondness of the human-built environment."[80] We understand this, without any significant problems, even without determining an essential meaning of the term *nature*. Indeed, it is the use of the term in contexts such as this—I stand outside, gesture to the trees surrounding my house, and say "I love nature"—that creates the meaning of the term.

Keeling contrasts this with several "odd" uses of the term *nature*, as if a person showed us a photograph of Times Square in New York City while stating that "I do nature photography" or if a person pointed to a computer while stating "it is amazing what nature can do." Although the words in these sentences make sense, we would be unsure what the speaker meant, for the speaker appears to be using the word *nature* incorrectly. "Cases like this demonstrate that there is an internal grammatical relation between human artifacts and nature or natural objects that cannot be genuinely doubted."[81] And the key purpose behind the use of the words *nature* and *artifact* in our "language-game" is to make a distinction between human agency and nonhuman agency.[82]

Because the terms *nature* and *artifact* have an internal grammatical relation, we cannot define them in some pure way independently of each other: "differentiating artifacts from natural objects is partly constitutive of the meaning of the two terms."[83] The distinction, and the use of the distinction to label some objects as artifacts and some as natural, is not open to empirical investigation. Here Keeling criticizes Vogel's question about human actions being different from nonhuman actions—"why are *those* processes called natural ones while the ones *we* initiate are not?"—as akin to asking why is black darker than white? For Keeling, "there is no justification beyond simply saying, 'we play this language-game, and *this* is how we play it.' There is no way to justify empirically the fact that human artifacts are not natural objects. It is true *a priori*."[84] The dualism of nature and artifact thus does not need to be defended; it is pre-supposed in any discussion of the value of the natural environment.

Chapter Three

So the critics of the wilderness idea—those that deny the existence of a nature free of human interference—are making an empirical and ontological claim about terms that are fundamental to our grammar, our language for describing the world. To say that empirically there is no place on Earth that is not untouched by human activity may be factually correct, but this does nothing to change our use of the terms *nature* and *artifact*. It does not demonstrate the truth of an anti-dualist position regarding humanity and nature. Nor can one reject dualism by changing the context of the word *wild* as Vogel does, in his use of the term to apply to human action. As Keeling argues against Vogel, "to extend the concept of wildness to the unpredictability of human artifacts … is not to make any new empirical observations about human artifacts or to discover any hitherto unnoticed facts about them. It is … simply to invent a new context for the word 'wild' where there are no established rules for its use."[85] This new use of the term *wild* makes no sense within our established grammar. We cannot dismiss the dualism of nature and humanly created artifacts by linguistic fiat.

This focus on the language we use in developing a normative theory about the value of natural entities is given additional support by a similar argument about the use of metaphors in debates over environmental policy. Willis Jenkins has argued that various descriptions of nature are really proxies for ideas about human behavior and action regarding the natural environment, so that we need "to pay evaluative attention to metaphors of agency."[86] To cite some obvious examples mentioned by Jenkins, if we use a metaphor of "raping nature" through human action we will have different ideas about environmental policy than if we use the metaphor of the "management" of natural processes. Thus, "we cannot suppose to begin ethics apart from the way roles and practices are already imagined."[87] Jenkins uses this focus on metaphors of human agency to re-locate the dualism of nature and artifact that permeates my critique of restoration. According to Jenkins, the dualism is not in my ontological "classifications of reality" but in my approval of just two extreme metaphors of agency—either we can preserve the integrity of nature by letting it be, or we violate it by acting and interfering with natural systems. It is the "limited conception of environmental practices" found in my arguments that "reinforces" the dualism.[88] Jenkins' solution, at least in part, is to develop a richer and more inclusive metaphor of

Ecological Restoration and Domination

human agency, for the restricted senses of agency that he claims to find in my argument actually interfere with the more complex view of nature and artifact that is necessary for a meaningful environmental ethic. The first condition for a proper metaphor of human agency is that "the concept must be able to accommodate various forms of the "natural" and complex gradations of "artificial," which is to say that it must be able to account for a rich variation of environmental particularity."[89]

So despite Jenkins' criticism of my too restrictive dualism of human agency regarding the natural environment, his conclusion is that we must develop language appropriate to a complex and nuanced view of artifacts and nature. This conclusion supports the analysis and argument of Keeling concerning the grammar of nature and artifact. In short, there is nothing incorrect about the dualism of nature and humanity that lies at the heart of my criticism of ecological restoration. On the contrary, this dualism is a necessary requirement for any meaningful discussion of environmental policy and ethics. As Val Plumwood explains, "without some distinction between nature and culture, or between humans and nature, it becomes very difficult to present any defense against the total humanization of the world."[90]

Nevertheless, there is a danger in relying too much on arguments concerning the analysis of language. As Kate Soper succinctly comments: "it is not language that has a hole in its ozone layer."[91] There is a reason that we need to make an ontological commitment, and ontological distinctions, to a nature that exists outside the realm of human activity. That reason is the actual existence of a real other world, the world of nonhuman natural processes. This is the world that we, as environmentalists, wish to preserve and protect. Soper again: "it is true that we can make no distinction between the 'reality' of nature and its cultural representation that is not itself conceptual, but this does not justify the conclusion that there is no ontological distinction between the ideas we have of nature and that which the ideas are about."[92] Our language signifies a real thing, nature, which is actually distinct from human cultural activity.

What we need, then, is a critical realism that accepts the ontological existence of a nature that is distinct from human activity but at the same time acknowledges the influence of our language and cultural constructions on our understanding of this other realm. One component of this critical realism might be a naturalistic account of the nature/culture dualism. Paul Moriarty has presented such an account, by defining culture (following J. T. Bonner) as "information transmitted non-genetically (or as the transfer of information by non-genetic means)." This permits a negative definition of nature as "that which is not a product of human culture."[93] With these definitions, we have a naturalized account of both culture and nature that incorporates a dualism without denying naturalism. Why is this important? As Moriarty argues, a dualism of nature and culture that in itself is naturalistic is necessary for a coherent understanding of Darwinian science. After all, Darwin's concept of natural selection as the process by which evolution occurs is meant to be distinguished from artificial (or human-induced) selection, as in the breeding process of domestic animals and plants. Moriarty concludes: "the denial of the nature/culture distinction is truly anti-Darwinian because it fails to understand the meaning of *natural selection*."[94] Moreover, Moriarty can use this naturalized definition of culture to distinguish human artifactual creations from those of the animal world: although it is true that animals also pass on information through non-genetic means, "human culture is unique in terms of the amount and kind of information we are able to accumulate and pass on from generation to generation and in the ways we are able to use that information to restructure the environment."[95] This naturalized account of the dualism thus avoids the main critical objections raised against the use of the human/nature or artifact/natural entity distinction.

But these arguments concerning the language of the human/nature distinction also point in a positive direction towards what is really at stake in debates over dualism and the critique of ecological restoration. As with Soper's warning about the hole in the ozone layer, the importance of recognizing the dualism is that it presents us with the ontological reality of a nature we wish to protect. Remember that Keeling claimed that the critics of wilderness preservation who based their objections on the existence of a pernicious and meaningless dualism were pursuing a red herring. The real issue, for Keeling, and for Soper and Moriarty—and for myself—is determining the value of a realm that is "other" than humanity. To deny the existence of this realm distinct from human action is to play havoc with our language, science, and conceptual framework

Ecological Restoration and Domination

for the world. But more importantly, it is to deny the existence of values recognized by all environmentalists, the values of the natural world.

A critique of the dualism of humanity and nature is, quite simply, a waste of time and effort. As Soper's broad survey of ideas about nature demonstrates, it is the political consequences and policies that are derived from our views of nature that are the main issue. "Nature," she writes, "does not enforce a politics."[96] There are good reasons for believing in the distinctiveness of humans and human culture. "The human predicament is sufficiently different from that of any other living creature to make it implausible to suppose that metaphysical naturalism is the automatic ally of ecology, dualism ... its obvious enemy."[97] What matters is not dualism or non-dualism per se, for "the commitment to either may be said to be less critical to the practices of the Green Movement than the evaluative interpretations that are brought to these different perspectives on the nature-culture, nature-humanity divides."[98] In short, it is how we use the distinction between humanity and nature—a distinction that our language and conceptual frameworks of the world will not permit us to ignore—that will determine appropriate environmental policies.

V.

I have argued that the recognition of the human-nature dualism provides a solid reason for rejecting the project of ecological restoration, a policy that encourages the total humanization of the natural world. Understanding the significance of the dualism of humanity and nature reveals the essential artifactuality of the products of the restoration process, even as we understand this dualism of nature and artifactuality along a spectrum, so that the degree of artifactuality can be more or less extreme. A critique of the restoration project maintains the environmentalist value in the "otherness" of nature, a realm that remains conceptually distinct from the human world even as it undergoes more and more anthropogenic modifications. A belief in the dualism of humanity and nature is thus not the problem; it is, rather, the solution, the means to preserve the value of the natural environment. It is the belief that can be used, as Anne Frank does, to resist the all-encompassing power of human evil and domination.

Let me conclude this chapter with some brief thoughts on the implications of this conclusion for actual and potential restoration activities—and indeed for some preservationist activities that intersect with ecological restoration. In the end I need to reconcile my thoroughgoing criticism of ecological restoration in general with my approbation of a limited beach replenishment project for the coastal communities ravaged by Hurricane Sandy. First, consider the fact that the maintenance of preserved areas requires human action. Although a strict preservationist attitude will prohibit the direct management of natural processes in a preserved area, banning the use of controlled burns or the culling of certain animals, even a total hands-off policy requires the creation of a boundary area, a borderline, a barrier to prevent intrusion from humans who may want to use the area. For this reason, Thomas Birch argued that to a certain extent even a "pure" wilderness area is an artifact of human production and power.[99] National monuments that are wilderness areas, such as the Giant Sequoias, would be, according to Birch, artifactual, for their continued existence requires the protection of human institutions. But note that if we employ the analysis of Siipi, discussed above, the evaluation becomes more complex. Using the first of Siipi's criteria, the Giant Sequoias, or any other natural wilderness entity or area, would not be an artifact, since the human action involved did not create the entity; but using the second criterion, one could argue that the human activity of setting up a boundary or a protective system, changed the function of the entity, at least in part, for now the protected entity has the additional function of being a symbol of a wild nature. Clearly this is where the emphasis on the spectrum of naturalness and artifactuality becomes extremely important. The artifactuality of protected wilderness areas or preserved national monuments is extremely small, falling at the end of the spectrum closest to "completely natural," as long as there is no direct activity that tends to preserve the natural entity. If the Forest Service, for example, chooses to allow controlled burns—or takes the opposite position of doing everything it can to prevent all forest fires in the area—then this human activity increases the artifactuality of the area. This artifactuality does not mean that the policy is evil, and it does not mean that the actions should be prohibited: the point is simply that we recognize the human influence in the continued existence of the natural area.

What does this mean for the policy of ecological restoration? I make no blanket condemnation of restoration. Even in "The Big Lie" I

Ecological Restoration and Domination

compared it to the cleaning or covering up of a stain on a carpet, an action that might be necessary to make one's living room presentable—but I claimed that far better would be the policy of preventing the stain in the first place.[100] So restoration projects are often better than nothing, but what we must always remember is that these activities are generally on the far side of the spectrum, near the extreme of artifactuality. Consider the restoration of abandoned farms in the American prairie. Is the controlled burn of these farmlands justifiable, so that the seeds of original prairie grasses can be reactivated? Or should the abandoned farmland just remain as it is, waiting centuries perhaps for nature to take its course? As a philosophical pragmatist, I must admit that any decision will depend on the specific piece of land, the actual situation at hand. Whatever we do, controlled burn-restoration or letting be, we will, in a sense, be imposing a human intention on the landscape as it now exists. This case fits precisely into the criteria of artifactuality developed by Siipi: restoration of the farmland to return to a prairie landscape will bring into existence a new entity or ecosystem with a new and different function. The prairie environment will result in different outcomes than the abandoned farmland. We will be creating a landscape that we humans wish to see in the world. Thus, the end result of restoration projects might be a more pleasing world, and even a better world, but it will be a world that reveals the imprint of human intentionality and design.

And so I return to the eroded beach of my Fire Island community. Any plan to re-build the dunes will be on the spectrum of artifactuality and naturalness within the restoration process. But the evaluation of the project will depend on the precise details of the technological activities that will be used. The building of a sea wall, for example, will be different than simply adding snow fencing to catch sand. A flimsy row of snow fencing does not create a new physical landscape, since it merely accretes sand in and around the spaced wooden slats. A similar strategic response might be the planting of dune grass. Yet a stone or steel sea wall significantly alters the landscape by establishing a permanent feature designed to change the flow of water and sand in a dramatic manner; it is a literally new environment. Thus unlike the slatted snow fencing, the construction of the sea wall establishes a new entity with a new function—the two major criteria of artifactuality in Siipi's taxonomy. Thus we need to limit the scope of the technological intervention in the re-building of the sand dunes; we need to preserve the most natural landscape possible. We need to keep our gaze on the existence of that part of nature that lies independent of human interventions.

Although we can debate the precise mechanics of the beach restoration project, the fundamental philosophical issue in the critique of ecological restoration does not concern policy. Ecological restorations will continue no matter what philosophical critics say in academic journals and books. We can hope—as Light and Higgs suggest—that the restoration projects will be done in the proper spirit of co-operation and respect for nature and human community. Ultimately, however, I believe that how restoration projects are done, and for what purpose, and under what conditions, is irrelevant to the fundamental question. The issue is not what we do. It is what our actions mean. Ecological restoration will always be an expression of the human project of the domination of nature, the attempt to control the world that is distinct and separate from humanity. To limit the scope of this domination we must maintain an ideal of a natural world free and independent of humanity.

Notes to Chapter Three

1. Katz 1995.
2. For all the relevant works, see Katz 1991; 1992a; 1992b; 1993; 1995; 1996a; 1996b; 1997; 2000; 2002; 2009; 2010; 2011; 2012.
3. Elliot 1982, 81–93.
4. Elliot 1982, 84–85.
5. Elliot 1982, 85–86.
6. Elliot 1982, 85–89.
7. This is the chief argument in Richard Sylvan's criticism of my argument. See Sylvan 1994, 48–78. Elliot modified his position to consider these kinds of natural restorations in his book-length treatment of the subject: Elliot 1997. I discuss Sylvan's arguments below.
8. Physical artworks, of course, are continually being modified in minor and major ways, from cleaning the surface of a painting to making repairs to canvas and plaster surfaces. Modifications and repairs to artwork exist along a spectrum—as I argue throughout this book that modifications and interventions to natural processes also exist along a spectrum. Context and the extent of work matter. It is one thing to clean and repair a Vermeer; it is quite another to re-paint a portion of the artwork. See the controversy over the "botched" restoration of the portrait of the Ecce Homo fresco in the Spanish city of Borja in the summer of 2012 (See *The New York Times* August 24, 2012) or the debate over the extent of the restoration of the ceiling of the Sistine Chapel.

9. The public controversy began with a letter sent to *The New York Times* by the famous Broadway composer Stephen Sondheim. See "Arts Beat", 2011. Also see a follow-up article, Healy 2011.
10. Light 2000, 54.
11. Light 2000, 62.
12. Light 2000, 64–65.
13. Light 2000, 67.
14. Higgs 2003, 186.
15. Higgs 2003, 194–95.
16. Higgs 2003, 214.
17. Hettinger 2012, 39.
18. Hettinger 2012, 40; original emphasis.
19. Higgs 2003, 236–37 and ff.
20. Higgs 2003, 239.
21. Sylvan 1994.
22. Lo 1999; and Vogel 2002; 2003.
23. Lo 1999, 253.
24. What this objection and response demonstrates, however, is that the notion of "intention" needs to be clarified. See, for example, an argument by Mark A. Michael (Michael 2001, 150–154). Michael, while broadly sympathetic to my account of the "artifactual" regarding human interference with natural systems, notes that an intention will have different normative values depending on how it is described.
25. Lo 1999, 254.
26. Ibid.
27. See Katz 2000, 41–42.
28. As Haraway notes: "By the late twentieth century, our time, a mythic time, we are all chimeras, theorized and fabricated hybrids of machine and organism; in short, we are cyborgs" (Haraway 1991, 150). But note that Haraway uses the image of the cyborg to break apart the dualisms of dominance; her goal is different than mine in this particular argument.
29. Lo 1999, 257.
30. Lo 1999, 258.
31. Vogel 2003, 150–51.
32. Vogel 2003, 152.
33. Brennan's original argument is in Brennan 1988, 88–91. I use Brennan's argument first in "The Big Lie," (Katz 1992a, 239); reprinted in *Nature as Subject* (Katz 1997, 104).
34. Vogel 2003, 153.
35. Vogel 2003, 157.
36. Vogel 2003, 155, original emphasis.
37. Vogel 2003, 157, original emphasis.
38. Vogel 2003, 158.
39. Katz 1993, 223–24; reprinted in Katz 1997, 122.

Chapter Three

40. It is possible that part of the reason for the child's existence might be the purposes of the parents: another hand on the farm; a royal heir; to "save a marriage." Neither Vogel nor I consider these cases to be significant to the argument.
41. I first used the work of Packard in my version of "The Big Lie" that appeared in *Restoration and Management Notes*; see Katz 1991, 93; reprinted in Katz 1997, 100–101. See also Packard 1988, 13–22. Vogel is not citing this article by Packard.
42. Quoted in Vogel 2003, 159.
43. Indeed, recent technological developments seem to suggest that in the not-too-distant future, human offspring will be more and more designed, as are artifacts. Parents may be able to choose the sex, eye-color, and other physical characteristics.
44. Lo 1999, 265–66.
45. Vogel 2003, 162, original emphasis.
46. Vogel 2003, 163, original emphasis.
47. Ibid., original emphasis.
48. Eliot 1925.
49. Vogel 2003, 167–68.
50. Vogel 2003, 150.
51. Katz 1995, 284; reprinted in Katz 1997, 145.
52. Vogel 2003, 167.
53. Compare an argument by Mark A. Michael (Michael 2005, 49–66) that shows how the idea that "humans are natural" leads to anti-preservationist environmental policies, unless we add normative content to the concept of "natural."
54. Vogel 2002, 32.
55. Vogel 2002, 35.
56. Vogel 2002, 38, original emphasis.
57. Siipi 2008, 74.
58. Siipi 2008, 75–76.
59. Siipi 2008, 77–78.
60. Siipi 2008, 79.
61. Siipi 2008, 95.
62. Siipi 2003, 415.
63. Siipi 2003, 415–16.
64. Siipi 2003, 417.
65. Siipi 2003, 418.
66. Siipi 2003, 419–20.
67. Siipi 2003, 420.
68. Siipi 2003, 424.
69. Siipi 2003, 424–25.
70. Siipi 2003, 425.
71. Siipi 2003, 425–26.
72. Soper 1995, 15.
73. Soper 1995, 38.
74. Soper 1995, 39.
75. Soper 1995, 40.
76. Soper 1995, 41.
77. Keeling 2008, 506.

Ecological Restoration and Domination

78. Keeling 2008, 507.
79. Keeling 2008, 508.
80. Keeling 2008, 509.
81. Keeling 2008, 510.
82. Keeling 2008, 511.
83. Ibid.
84. Ibid.
85. Keeling 2008, 512.
86. Jenkins 2005, 136.
87. Jenkins 2005, 143.
88. Jenkins 2005, 146.
89. Jenkins 2005, 147, emphasis removed.
90. Plumwood 1998, 676.
91. Soper 1995, 151.
92. Ibid.
93. Moriarty 2007, 237.
94. Moriarty 2007, 242, original emphasis.
95. Moriarty 2007, 239.
96. Soper 1995, 141.
97. Soper 1995, 174.
98. Soper 1995, 176.
99. Birch 1990, 3–26.
100. Katz 1992a, 240; reprinted in Katz 1997, 106.

~ Chapter Four ~

INDEPENDENT NATURE DENIED

I.

I have argued that ecological restoration is an expression of the domination of nature. I have also claimed that the technological domination of humanity is a source of the evils of the human social and geopolitical world that encompass Anne Frank, her family, and all the victims of Nazi oppression. A central topic of this book is the analysis of several forms of domination and how these forms were bound together in the example of Nazi ideology and practice. A technological system that is embedded with the evil values of oppression and domination stands opposed to the autonomous unfolding of both a liberated nature and a free humanity. Anne's tree, once again, stands as a symbol of an independent nature that is used by her—and can be used by us—as an ideal of resistance to the evils of domination, a source of hope for a harmonious and peaceful world.

In Chapter Five, I will demonstrate in more detail the close connections between the process of ecological restoration (which I criticized in Chapter Three) and the Nazi system of oppression and domination as it was manifested in the environmental policies of the Third Reich. But in this chapter I need to deal with a serious objection to the foundations of my position regarding the independence of nature. One might argue that my view (and Anne's) is based on the reality of an independent nature that exists external to human influence and control, but that in fact, no such nature exists. We have already seen (and answered) one version of this objection in Chapter Three: the argument of Steven Vogel, where Vogel claimed that because of the pervasive influence of human activity on the natural world, no unmodified natural entities remain. Vogel used this claim in an attempt to undermine my criticism of the process of ecological restoration, for if the entire natural world has already been modified by human activity (and technology), then there is no special reason to avoid or prohibit projects of ecological restoration. The implication of this argument—a result that I do not believe is ever expressed or supported

Independent Nature Denied

by Vogel himself—would be a denial of the need for the protection of the natural world. If no independent nature exists, then why must we act to preserve this nature? Why not use our science and technology to manage, manipulate, and control all so-called natural processes for the best results of humanity? Since we already live in a modified world that is essentially artifactual, humanity should make decisions regarding the so-called natural world based on the promotion of human good. There is no reason to worry about the wholesale domination of nature because an independent nature no longer exists.

The objection of Vogel is primarily based on a factual claim: the actual physical modifications of the natural world by human activity. As such, it is similar to (and is based on) the argument made by Bill McKibben in his book that warned of the threat of global warming, *The End of Nature*.[1] McKibben claimed that since human science, technology, agriculture, and industry had altered the earth's atmosphere, there was no independent nature remaining. Modifying the atmosphere was equivalent to modifying everything. In this chapter, I do not respond to this factual argument. I believe that my arguments in Chapter Three concerning the spectrum of naturalness and artifactuality are sufficient to deal with this empirical claim: although the entire natural world has been altered by human activity, there are degrees and levels of change and modification. Just because the atmosphere is different today than it was in (let us say) A.D. 1200, this is no reason not to appreciate the nearly pristine nature of the great American wilderness preserves. And it is surely no reason not to continue to preserve and protect these wilderness areas: they are different from forest plantations and parking lots. So my concern in this chapter is not this factual or empirical objection to the reality of an independent nature. I am more concerned with a conceptual argument, one that is based on the general idea that nature itself is a construction of human activity. We can call this objection the "social constructivist" critique of the existence of an independent nature.

As a prelude to an examination of this objection, it must be noted that Anne's tree itself is not an actual example of a pure and independent nature. It is an urban tree, growing in the backyard of an Amsterdam townhouse. We can assume that it was planted and tended by humans who wished to increase the beauty of their urban domestic home. As in the process of ecological restoration, the tree is the result of a human intention to improve the world. Does this make a difference in how we should view the significance of the tree? I think not. Remember that for Anne, viewed as a writer and thinker, the tree serves as the opening move in a subtle argument concerning the importance of a nature that exists outside the realm of human evil. From the very first mention of the tree, she connects it to various other parts of the natural world—the sky, the moon, and birds—all of which are clearly beyond the power of human manipulation. Thus although the actual tree is clearly a domesticated product of human power, it serves as a powerful metaphor—for me but especially for Anne—of the value of a natural world independent of human influence and control. I will return to this point at the conclusion of this chapter.

So what does the social construction of nature mean? Vogel provides an introductory summary,[2] with the central point that since nature must be known to us through human categories of thought, there is no essential nature beyond what is constructed by humanity. He starts with poststructuralism's critique of the appeal to a foundation, origin, or immediate experience in human thought and knowledge claims. Language and/or social processes mediate all that we believe, and there is "an imperative to uncover, within everything that appears to be given, immediate, foundational—in a word, 'natural'—the hidden processes of construction and mediation that produce that appearance." This deconstructionist imperative leads to a "critique of nature," for much of environmental philosophy and policy uses the idea of nature as a foundation for explicating and justifying the norms and goals of behavior. But how is this use of nature—as a foundation for goals and norms—possible? Such a use of nature would seem to require a direct knowledge of what nature is, but this is impossible: "nature always appears to us mediated through language, concepts, worldviews, and personal and social histories that are particular and contingent; it never appears nor could it appear as it is 'in itself.'" There is no nature, Vogel concludes, beyond what is structured by human thought, language, culture, and praxis. There is no pure and immediate or essential nature that can be the standard of normative claims.

I used a pragmatic argument to deal with this objection in Chapter One, where I argued that we do not require the knowledge of nature in itself, beyond the existence of any social categories, in order to know

Independent Nature Denied

what the oppression, domination, and, ultimately, the liberation of nature is. My argument was based on the comparison with human oppression and liberation, for although we do not know what human nature is in itself, we can still make practical judgments regarding human domination and its opposite. Here I want to go beyond that earlier pragmatic response to look more deeply into the nature, so to speak, of the social construction of nature.

Vogel considers two primary ways in which humanity constructs nature. The first is the process of science itself.[3] Contemporary studies in the philosophy of science clearly point to the fact that scientific theories—and indeed, the objects of scientific thought—are products of a complex set of specific social practices. The objects and theories of science are essentially artifacts, produced in science laboratories by means of a precise set of rules that confirm legitimacy and truth to the knowledge of nature that is the result. We might want to say that biology, chemistry, and physics discover the facts of nature, but what the sociology of science demonstrates is that nature "is something to which we have access only through the practically and socially organized activity of scientists."[4] Scientific practices transform our experiences of the natural world into a formal system of observations. So facts about nature are not *discovered* so much as they are *made* in the organized procedures of science. Experiences of nature that do not undergo the rigor of scientific practices are not considered to be real or valid. The scientific process is a construction of the reality of natural entities, for only entities that pass through the organized system of science are considered to be real.

But Vogel also emphasizes a second process of construction, not limited to the discourse and thought of science: the literal construction of nature and the world by human activity.[5] Humans are in the world, according to Vogel, as active transformers. Everything that we humans do physically shapes the environment in which we are embedded. Physical nature has already been humanized through countless millennia of human activity. We see this easily enough in the developed world, where none of us actually lives in a natural landscape—even the Fire Island home that I discuss in Chapters One and Three is located on a piece of land heavily modified by the human technology of water channels, bulkheading, and beach replenishment. But our romantic view of the pristine and independent nature of the undeveloped regions of the third

Chapter Four

world, according to Vogel, is based on a "traditional (and racist) dualism that tends to relegate indigenous populations ... to the category of the natural, and hence the nonhuman."[6] We tend to overlook the ways in which even so-called "primitive" people have actively transformed the natural habitats in which they live. We accept the areas of the undeveloped world as more natural because we do not consider the modifications made by indigenous humans as *human* modifications. We fail to see that nature has been made—constructed—into a human landscape.

Vogel's conclusion, again, is that no nature independent of human activity actually exists because nature has been socially constructed by human knowledge and praxis; thus an independent nature cannot be used as a guide to norms, values, and goals of human behavior, and especially not as a guide to environmental policy. (We saw how some of the implications of that conclusion played out in the debate over the policy of ecological restoration in Chapter Three.) Not surprisingly, some environmental philosophers, even those sympathetic to the claims of poststructuralism, have attempted to weaken or overturn this conclusion, to show that the idea of the social construction of nature need not be damaging to the environmentalist ideals of preservation and conservation.

Anna Peterson begins by emphasizing the difference between two versions of constructivism, which are roughly equal to the two methods of constructing nature that we saw in Vogel. On the first version, nature is constructed in the processes of human thought: "different individuals, times, and societies construct particular versions of nature insofar as they interpret it in different ways in and through cultural categories and values."[7] Our interpretation and knowledge of nature—including science and discourse—is culturally or socially dependent. A second version of constructivism is more extreme: it is the physical construction of the environment, or what Vogel called the literal construction of nature. In this version, all landscapes have been modified by human activity; even wilderness areas have been transformed by the effects of anthropogenic climate change.

For Peterson, there are ethical "dangers and possibilities" in both versions of constructivism. In the extreme literal version of constructivism, there is a fear that there will be no reason to protect natural entities, such as species, wilderness areas, and uncontrolled natural systems such as free-flowing rivers[8]—what I have called in this book the autonomous

Independent Nature Denied

processes of nature. If humans have constructed all environments, then there is no nature—wilderness, species, autonomous systems, etc.—that needs to be protected. Concerning the "softer" version of constructivism, she uses an argument that I used against Vogel (in Chapter Three), noting the extreme relativism of the position: if all of our ideas about nature are subject to cultural interpretation, then we will have no standard to evaluate one constructed environment as better than another. Indeed, if all of nature is the result of human modifications, and all modifications are open to individual and cultural values, then we will have no criteria for judging what interventions in nature are good or evil.[9]

And yet Peterson does not deny the existence of constructivism, and she sees that there are possibilities for conclusions favorable to environmentalist ideals within its claims. One interesting possibility is the idea that social constructivism can challenge the dominant anthropocentric vision of the human relationship to the natural world that is prevalent in Western civilization. (To be fair, Peterson also thinks that the dominant vision of a nature-humanity dualism can be challenged, but, as I have argued in Chapter Three, I believe that such a dualism is essential to a meaningful environmental ethic.) Social constructivism can challenge anthropocentrism first by taking seriously the interpretations of nature that are seen in other non-Western cultures. Peterson cites two indigenous populations that make no distinction between certain animal species and humans: their interpretation of the rough equality of humans and nonhumans is no more nor less a construction of nature than the dominant Western view of human superiority.[10] And this leads to an even stronger claim about the nonanthropocentric construction of nature; for animal species, as well as humans, also physically transform their environment and habitat. Nonhumans also construct the world. This could lead to the ethical conclusion that we should respect and take seriously not only the interpretations of the world by other humans but also the actual habitats shaped and modified by both humans *and* nonhumans.[11] Here the physical construction of nature by natural nonhuman entities supports the conclusion that we should respect the autonomous processes of nature—as I argued in Chapter One, we should, as far as possible, let nature be.

Peterson's analysis shows that neither form of social constructivism provides a fatal objection to the environmentalist ideals of preserving

Chapter Four

and conserving a natural world distinct from human society and culture; instead there may be possible ways that environmentalists can use the constructivist platform. Mick Smith takes this argument even further, arguing that environmentalists do not need to accept the scientific naturalism that is usually opposed to social constructivism, because the ontological issues surrounding the reality and character of nature are less important than issues of value-determination.[12] Smith begins his argument with a taxonomy of various positions in social constructivism that is more complicated than the summaries presented by either Peterson or Vogel, but I need not review those positions here.[13] For my purposes, it is sufficient to note that Smith also makes the central distinction between a social constructivism that focuses on the role of human thought to create various interpretations of nature and a social constructivism that claims that nature is actually constructed by human activity. The former version of social construction would involve the social construction of our knowledge of nature, and thus can be called epistemological constructivism; the latter is an ontological thesis based on the idea that nature's being is merely a product of human action.[14]

For Smith, the importance of the distinction between epistemological and ontological constructivism is that it permits us to separate the two kinds of issues concerning the essence of nature, issues concerning knowledge of nature and issues concerning the reality of nature. Indeed, we should bracket out all of the questions about the ontology of nature, for social construction as a theory regards such claims as unknowable. "The constructivist approach is generally to *suspend*, rather than make, claims about the world's ontology, since these kinds of *claims* are, rightly or wrongly, regarded as culturally bound and hence ultimately undecidable."[15] Once we move away from issues of the being or ontology of nature, we can find the importance of social constructivism in the realm of value determination. For Smith, the issue becomes "one of *location*, of where values are produced and their degree of attachment to that locus of production.[16] Different positions regarding the source and location of the value of nature will be developed. Some forms of constructivism—what Smith calls "strict constructivism"—will claim that values are "constituted within the symbolic, ideological, and political order of society" as cultural products. Nature is thus merely "a 'sign' with shifting patterns of meaning" determined by its relation to other signs.[17] As one

Independent Nature Denied

alternative, Marxist-based theories will claim that values are produced by the economic relations within a society, for "the cultural superstructure's relation to nature-in-itself is always mediated via the economic base."[18] Then again, deep ecologists and other environmental theorists who tend to support the existence of nonanthropocentric intrinsic value will use the autonomous processes of "nature" as the source and location of value. Nature is thought to be a self-generating "productive field" and each of the entities within nature is thought to have its own internal good.[19]

Smith claims that "there are no fundamental ontological differences between these positions."[20] Rather than making ontological claims about the reality of nature (or the nature of reality!), these theories are merely offering "differing analyses of social/natural relations."[21] Using an argument similar to the one I used in Chapter One, Smith claims that there is no need to get to the ontological foundation of "nature" in order to use the concept in the determination of values—but there is also no need to get to the ontological foundations of "society" or "economy" or "culture." All are the products of human thought. Conceived as the products of social construction, they can all be used in the determination of our values. The key point, however, is that "nature" is "part of the *context* constructivists must look to."[22] We can reject a naturalistic ontological claim that there is one master narrative that reveals (through science) the true reality of nature while at the same time realizing that "nature" as a social construct is no better or no worse than "society" or "culture" as a location of our values. "Nature is indeed contested, but it is also a contestant, a constitutive part of the medium of our existence."[23] As such a contestant, nature produces value as surely as human culture does. Thus, for Smith, "there is nothing nonsensical in valuing those parts of 'nature' which we choose, or are brought, to recognize either through our own or nature's activities."[24] In short, it is possible to believe in the value of an autonomous system of nature even within the perspective of social constructivism. Although we can acknowledge the validity of social constructivism regarding nature, we need not believe that we humans are "the *be all and end all* of the world"—we need to recognize that there is value in what is produced by the nonhuman world of nature.[25]

In sum, the thesis that nature is a social construction does not serve to invalidate the overall argument of this book that a free and autonomous system of natural processes and entities can be used a source of hope and resistance to the forces of domination. The nature that is conceived as independent of human action may be a social construction, yet it remains a necessary part of the context of our normative claims about the human presence in the world. In the next section, we will look more deeply at this context—the intersections of history, culture, and nature—as we continue to examine the possibility and power of the idea of an independent, autonomous nature.

II.

The first two chapters of this book argued, in part, that we need a comprehensive understanding of both human history (including technological progress) and the processes of nature in order to develop an appropriate response to the evils of domination and oppression. Nature and history are inextricably linked, so that any attempt to seek values, goals, and norms must include an account of each of them, and indeed, an account of how they co-exist and mutually relate. The theory of social constructivism, however, raises a possible objection to this argument, for it can be understood to challenge the existence of an independent nature outside of human influence and control. Such a theory would suggest that only human history and culture is important, for nature is a mere product of the ongoing development of the ideas (including science) of human societies. In the previous section of this chapter, I presented a primarily theoretical argument that concluded that even if nature was a social construct, it still was part of the context—with human history and culture—in which we developed a system of values and norms. In this section I look closely at a social constructionist argument that is grounded in the historical and cultural context of the human relation to nature. It is thus an objection that fits squarely into the central methodology of this book, the intersection of human history and nature.

Although it does not engage the moral arguments I am considering, one of the more interesting attempts to express this social constructivist objection is found in historian Simon Schama's book, *Landscape and Memory*.[26] In this fascinating mixture of history, art criticism, cultural geography, and philosophy, Schama claims that "landscapes are culture before they are nature; constructs of the imagination projected onto wood and water and rock."[27] The metaphors we use to understand nature become the reality, more real than the actual entities themselves: such is

Independent Nature Denied

Schama's conclusion. I remain unconvinced by his argument, although I find his reflections on nature, culture, and history immensely valuable. His position is that nature and culture are indivisible: "landscape is the work of the mind. Its scenery is built up as much from strata of memory as from layers of rock."[28] And he purports to demonstrate this idea through a survey of the history of landscapes, of the human interaction with nature, through art, religion, war, and commerce.

The first section of Schama's book concerns forests, and is thus the most relevant to our discussion of the meaning and value of trees—and ultimately, our interpretation of Anne Frank's chestnut tree. The main idea is that different societies throughout history have had different relationships with the forest, different myths and understandings that determine the ways in which these societies interact with the forest landscape. Germans, for example, began to use the idea of the forest as the natural realm of their people as a way to differentiate themselves from the Roman invaders, who felt threatened both by the untamed forest wilderness and the wild barbarian people that lived there. By the sixteenth century, Germany's enemy—still based in Rome—was the Catholic Church. Now German geographers wanted to answer the southern European criticism of the northern forests as beastly and ugly—a view also expressed about German civilization; thus they extolled the many wonders of the German forest landscape.[29] By the end of the 1700s, the German philosopher Johann Gottfried Herder was criticizing the universal classicism based on the Greeks and Romans, and instead arguing for "authentic native culture" that would be "organically rooted in the topography, customs, and communities of the local native tradition."[30] Herder argued that the Middle Ages, not the Enlightenment, was the true source of the best of the German virtues, an "unspoiled native landscape" and a time that was "sacred, communal, and heroic": the forests and trees were "the emblem of Germania itself."[31] (We will return to the German fascination with an authentic native culture in Chapter Five.)

England and America also have their myths of the forest landscape. The English national narrative entails the individual freedom of the common citizen roaming the greenwood in opposition to the authoritarian rule of the nobility, personified most famously by the legends of Robin Hood and his men of Sherwood Forest. In America the myths of the forests begin with the Puritans and other original settlers of the New

Chapter Four

World who saw the untamed wilderness as an alien pagan place in need of civilizing. Yet by the nineteenth century, with industrial development gaining momentum, the need arose for the preservation of the original American landscape. Schama considers the discovery in 1852 of the Giant Sequoias in Yosemite Valley, California, as emblematic of the new (but still complex) American attitude. When first discovered, the trees were exploited as monumental and freakish tourist attractions, but eventually they become a symbol of the "holy asylum" of the Promised Land: "its foliage trickles with sunlight; its waters run sweet and clear. It is the tabernacle of liberty, ventilated by the breeze of holy freedom and suffused with the golden radiance of providential benediction."[32]

Throughout this historical exegesis, Schama is aware of the irony that attends the acceptance of these forest myths. "By the time the German forest was being identified as the authentically native German scenery [in the sixteenth century], much of it was fast disappearing under the axe."[33] Indeed, in each of the national and regional narratives Schama surveys, he makes the point that the intensity of the nature myths seems to increase just as the commercial and economic development of the forests begins to take hold. In the late 1500s, for example, "just at the time that Robin Hood's Sherwood was appearing in children's literature ... the greenwood idyll was disappearing into house beams, dye vats, ship timbers and iron forges." Because the forest was so important for both military and economic development, the guardian of the national forest "was bound to be torn between exploitation and conservation."[34] The debate has continued for five centuries, as Schama notes—through the arguments of John Muir and Gifford Pinchot in late nineteenth-century America about the values and uses of the national parks and all the way to the contemporary arguments about the meaning of sustainable development. My argument in this book, in part, can be considered to be a continuation of the debate, for it examines the meanings of the concepts of domination, oppression, liberation, and autonomy as they impact the human relationship to the natural world. Is nature to be used, managed, modified—in short, dominated by humans—or is it to be preserved, set free to pursue its autonomy? To borrow a phrase from Schama, does the national "bureaucrat ultimately [prevail] over the loose-blouse Romantic conservationist?"[35]

Independent Nature Denied

Although I do not find Schama's historical survey problematic, I begin to part ways with his philosophical analysis about its meaning. It is true that different national and regional narratives inform and determine differing relationships to local and regional natural landscapes. But Schama wants to go further, to claim that it is the human relationship to the landscape which defines the landscape, defines nature in its interaction with humanity, so that the reality of an independent nature is yet another myth or metaphor. An independent nature has no reality outside of the conscious meanings of humanity. Yet if these myths of a national forest character seem to arise during, and despite, times of the economic development of the forests, then it would seem that the myths themselves are not wholly determinate of the human relationship, and human meaning, of the forests. The myths and stories of a national forest character are being created as a means to impede the so-called "progress" of economic development, to impede a form of domination. The narratives do not determine the reality of the forests; rather the narratives are dependent upon—maybe even parasitic—of the actual independent existence and reality of the natural forest itself.

For me, Schama's view espouses another form of the human domination of nature: call this "epistemological domination." It is clearly a form of social constructivism, for it posits the idea that human thought and language determine the nature of reality. Epistemological domination is a very seductive idea, especially in an age, such as ours, that desires unlimited economic and social progress. For if our thought determines the reality that surrounds us, then we can control this reality with impunity, and make the world what we desire it to be. Epistemological domination leads to the actual physical domination of the natural world, for it can be used to justify a variety of oppressive environmental policies, from the development and destruction of pristine wilderness areas, to the management of forests as timber plantations, or to the ecological restoration of exploited landscapes. We can level forests for economic prosperity and military power, yet maintain the beautiful stories of a liberating wilderness (in Schama's words) "suffused with the golden radiance of providential benediction."[36] We can develop and destroy any ecosystem or natural landscape, secure in the knowledge that the technology of ecological restoration will return the landscape to its original form and substance. Epistemological domination is the doctrine that not only our

Chapter Four

science and technology can master nature, but our ideas can master it as well. This means that we can alter and manipulate reality while at the same time we can believe confidently that we are respecting the integrity of natural processes. Epistemological domination is thus the necessary condition for the justified abuse of nature: it enables us to achieve our anthropocentric goals while simultaneously oppressing, dominating, and destroying the natural world, while all the time believing that we are doing good. Here we see the primary threat of social constructivism to the preservation of the integrity and autonomy of natural processes.

But this form of the theory of the social construction of nature—epistemological domination—exhibits internal contradictions, and this can be seen not only in the theoretical arguments I presented in the previous section of this chapter, but also in Schama's own memoir regarding his grand history of forest landscapes. As the earlier chapters of this book demonstrate, I firmly believe in the power of personal experience to make a philosophical argument. So let us use Schama's own experiences to refute the claims of epistemological domination. It is more than appropriate that his personal narrative history concerns the Jews of Eastern Europe, the primary victims of the Holocaust. His book begins with a deeply moving story about his search for his family's Jewish roots in the forests of Lithuania and Poland. Yet one of his colleagues snidely remarks to him: "Trees have roots. Jews have legs."[37] Schama at first dismisses this hurtful comment, but when he discovers the remains of the Jewish cemetery at Punsk, the gravestones now covered with layers of soil and vegetation, he begins to see the truth. "The headstones that had been lovingly cut and carved were losing any sign that human hands had wrought them. They were becoming a geological layer ... as verdant Lithuania rose to reclaim them."[38] This passage is eerily similar to my description, in Chapter One, of the Warsaw Jewish cemetery—a description that I wrote originally in 1995, at exactly the same time that Schama wrote his.[39] Schama's experience (and the conclusion he draws from it) is the same as mine: he begins to realize the overwhelming power of an autonomous nature to re-assert itself against any and all human forces, institutions, and ideas. Once there was a Lithuania with no Jews (and no people)—just forests. And then there was a time when there were Jews: "and now there are no Jews again and the forest stands there."[40] Trees have roots. Jews have legs. (We mention without comment here

Independent Nature Denied

the fact that the Jewish legs did not move voluntarily out of Lithuania.) Schama walks away from the remains of the Jewish cemetery, now enveloped by the forest. A liberated nature is more powerful than human epistemological domination. The forest exists independently of human myth and metaphor. The forest that Schama experiences enveloping the headstones at Punsk is not a social construction.

III.

What is the value then in considering the theory of the social construction of nature? It is undoubtedly true that on the level of interpretation and knowledge claims human individuals and cultures determine the meaning of the natural world. As a prelude to a critique of social constructivism David Kidner writes, "few environmental writers would quarrel with the notion that our understandings of nature are affected by our cultural background, training, language, and so on, or that unmediated contact with nature is unrealistic."[41] I would go even further: the unmediated experience of nature is impossible. We live, after all, in a post-Kantian world; all of our knowledge claims, beliefs, and experiences are filtered through human categories of understanding. The problem is the idea of a literal or physical version of the social creation of nature: does our understanding and interpretation serve as the actual construction of physical reality?

Kidner argues that the theory of social constructivism itself is part of the process that I have called the domination of nature. He does not mean this in the sense of the "epistemological domination" that I discussed in relation to Schama's account of the history of forests, but in the actual domination of physical reality. Social constructivist arguments about the meaning of natural reality provide the conditions for the physical transformation of nature. As Kidner notes, "the intellectual dismemberment of reality is often a precursor to and a legitimation of its physical destruction, and academics as well as logging companies have contributed to the degradation of the natural world."[42] As I noted above, if the natural world is merely a creation of human categories of thought, then why not mold it into whatever shapes and forms that are most appealing to us? The increased industrialization of contemporary society through science and technology can lead to the direct modification and manipulation of the natural world. Nature, according to Kidner,

Chapter Four

was not constructed by human social categories such as language and science, but it is now being re-constructed by them.[43] The genetic revolution of agriculture and livestock species is perhaps the most obvious example. Ecological restoration would be another. And the modification of the human species—through processes such as recombinant DNA, genetic testing, or fetal surgery—would also be part of the intentional re-construction of the physical world. Thus Kidner concludes, "social constructionism, then, can be seen as rooted within a broader reconstructive project which reconfigures both humanity and the nonhuman world according to an industrial blueprint."[44]

Again it has to be emphasized that the physical construction or re-creation of the natural world is only possible because we have accepted the idea that our language, concepts, and science determine reality. Kidner writes: "What has happened here is that since we have lost touch with any frame broader than that defined by our language and our social 'reality,' anything beyond this 'reality' will necessarily seem unreal, invalid, or nonexistent."[45] The result, for Kidner, is a form of solipsism.[46] And the solution—and this is *my* solution, not Kidner's—is to reinforce the necessity of the human-nature dualism. We must acknowledge the existence and reality of a nonhuman world outside the realm of human language, thought, and science. Only by acknowledging the reality of an independent nature can we avoid the tendency to modify and manipulate the entire world—human and nonhuman—to meet our needs and desires.

We thus return to topic of dualism, first broached in the concluding section of Chapter Three. There I used primarily the work of both Kate Soper and Paul Keeling to argue that the dualism of humanity and nature is rooted in our language, discourse, and thought. For Soper, remember, "the a priori discrimination between humanity and 'nature' is implicit in all discussions of the relation between the two,"[47] so much that at least in Western thought the dualism "remains a presupposition of all its philosophical, scientific, moral, and aesthetic discourse."[48] Keeling introduced a version of this argument based on the philosophical ideas of Wittgenstein: the relationship between humanity and nature, between the concepts of the "natural" and the "artifactual" was one of grammar. "There is no way to justify empirically the fact that human artifacts are not natural objects"—it is implicit in the structure of our language-game.[49] If someone showed us a photograph of the New York

Independent Nature Denied

City skyline and told us that he loved "nature photography" we would not know what he meant—he would be making a fundamental error in grammar, in the meaning and relationship of words.[50]

It is ironic (perhaps), but this argument for the dualism of nature and humanity, which I introduce here to confront the tendency of the social constructivist position to lead to the physical domination of the natural world, can be considered itself to be an argument with roots in social constructivism. It is an argument for the existence of a human and nature dualism that is based on language. It is not an empirical argument; it is a claim about the way in which human discourse and thought structures reality, in this case, the way it structures the relationship between humanity and nature. So it is not social constructivism *per se* that I am challenging in this chapter (and in this book); rather, it is the use of the theory of social constructivism to deny the existence of the idea of a natural world independent of humanity. As I argued in Chapter Three, the existence of this dualism is necessary for any meaningful environmental policy that can protect and preserve the natural world.

Can an explicit social constructivist position support the dualism of humanity and nature? My reading of Neil Evernden's *The Social Creation of Nature* suggests that it can. Evernden traces the history of the idea of nature focusing on a central ambiguity, that nature is the external material stuff that is studied by science, or that nature is what is normal, the way that things are supposed to behave.[51] The first meaning of nature—nature as the material given of the world—"lends an aura of objectivity and permanence to the understanding of nature as norm."[52] But modern science requires the banishment of norms and values from the objective world that it studies. As it creates the world through the establishment of theories and principles, science must purify nature and eliminate all meaning and purpose, such as Aristotle's final causes[53]—the nature revealed by science is a realm of necessity and laws, not a realm of subjectivity or willing. The world of purpose, norms, and will is reserved for humanity. "The conceptual purity of the domain of Nature is a condition for the security of the realm of Humanity."[54]

Nature is a social creation because humanity needs a realm that is distinct from the realm of human freedom. "Nature, though explicitly nonhuman, is ours: we do not so much read the 'book of nature' as Galileo desired, as write it."[55] Evernden even calls "nature" an artifact, a tool that can be used by humanity for whatever purpose it desires. And its principal purpose is to be used to differentiate the world of humanity from the world of physical necessity. "For the humanist concept of 'Human' to exist [i.e., an idea of the human being as a free, autonomously willing creature with a moral purpose], we must first invent Nature: our freedom rests on the bondage of nature to the 'Laws' which we prescribe."[56]

Danger arises, however, in what might be called the imperialism of modern science: it continually attempts to conquer every realm of existence, including that of humanity. Evernden calls this the "modern monism" of science and in it nature tries to reassert its control over humanity. "Once we accept, through the study of Nature, that all life is organically related, organically the same through the linkage of evolution, then humanity is literally a part of nature."[57] We humans would no longer be free autonomous beings with wills that control our decisions, our activities, and our lives. We would be totally subject to the physical laws of the universe, just as any other nonhuman natural entity that we choose to study and know. The physical sciences would be able to analyze, organize, know, and control all aspects of human behavior and life. Our own science tells us that nature has reclaimed us as part of its system of necessity. How then, do we "get off our own dissecting table?" Only by admitting that "nature" is a social creation, admitting that the idea of an independent nature that operates under universal objective laws is a fiction that we have created in order to think of ourselves as qualitatively different from the nonhuman world. "We are going to have to admit our own role in the constitution of reality."[58]

Yet this conclusion is also unacceptable. It succumbs to a different form of monism: not the monistic materialism of modern science, but the monism of "modern idealism," for now it seems the entire world is just a system of humanly created ideas. This idealism, it seems, would lead to the solipsism of which Kidner warned: the belief that the entire world can be made amenable to our human desires, for the entire world is just our creation. In addition, the materialism underlying modern science is not going to be abandoned: it has been far too successful in providing the technology necessary for human life. So we require the dualism, the belief in a separation of humanity from nature, even though the principles and laws of nature launch a continual assault against it.[59] How do we withstand the assault? Evernden enlists an idea of Maurice Merleau-

Independent Nature Denied

Ponty that we "return to things themselves" before they become a part of the human system of knowledge. "Once named and explained, they become social creations, and their primordial givenness is subordinated to their social utility."[60]

Evernden gives an example that will remind us of Schama's cultural history of the forest:

> A forest may be a mythical realm or a stock of unused lumber, but either way, it is able to serve a social function. It is, in that sense, never *itself* but always *ours*, our "system" of distinctions among the worldly phenomena.[61]

The only way to see the forest for itself is to attempt to experience the contents of the forest before the imposition of human explanation and knowledge. We need to attain an understanding of what Evernden calls the "ultrahuman"[62]—that which lies outside human life and thought. If we do not accomplish this, we will be trapped forever in a world of human creations, a world of human technology and human artifacts. And then we will become artifacts too: "If we can only look to cultural artifacts during our generative process [and that is all that scientific knowledge can give us], we must become, in a sense, cybernetic beings, creations of our own technology."[63]

This is the exact same danger that I identified in my discussion of the practice of restoration ecology: the creation of a totally humanized, artifactual world. And my solution, the belief in a dualism of nature and humanity, and the attempt to preserve the distinct realm of nonhuman natural processes, is the same as Evernden's. Even from the perspective of social constructivism, Evernden recognizes that we must experience the "otherness" of a natural world that is "wild"—self-willed and radically independent from humanity.[64] Once we seek to tame it, domesticate it, modify it, or explain it, the otherness will disappear and it will become a part of the human realm. Then the entire world will be a human world. Only by the direct experience of the wild otherness of nature can we hope to escape a world dominated by humanity.

IV.

The history of human culture teaches us that we can maintain the idea and experience of a natural world separate and distinct from humanity. This is the central theme of Robert Pogue Harrison's *Forests:*

Chapter Four

The Shadow of Civilization, a fascinating cultural history of the natural world, and in particular, forests.[65] Harrison is a professor of literature and his account of the human relationship to nature and forests draws heavily from a close reading of literary texts. His overall topic is "the role forests have played in the cultural imagination of the West,"[66] and his guiding theme is that "forests represent an outlying realm of opacity which has allowed … civilization to estrange itself…" In short, forests represent a world apart and different from human society. Because of that separateness humanity is able "to project into the forest's shadows its secret and innermost anxieties,"[67] but also, I would add, its hopes and aspirations. The distinct physical reality of the forest acquires meaning from the human cultural imagination.

Harrison's wide-ranging survey begins in antiquity and concludes with the end of the twentieth century. I will only note a few highlights that are relevant to my argument in this chapter and this book. The development of Rome for example, involved a complex interplay with the forests of Latium. In the founding myth of Rome, Romulus and Remus find safety in the forests as the asylum necessary to save their lives, but when Romulus founds his city of Rome it is an opening in the forest, and it becomes an asylum for those seeking the peace and serenity of civilization. The forests become literally a place of no one, outside the bounds of civic life.[68] In Harrison's view of this history, the city represents the triumph of reason and critical thinking over the chaos and disorder of the Dionysian realm—and the symbol of this triumph is Socrates, who not only vanquishes the Dionysian impulses of his companions in Plato's *Symposium* but also offers a criticism of rural life in the opening pages of Plato's *Phaedrus*.[69] The forest was a place to be feared; its laws and organizing principles were alien to human society. And this mythic meaning plays itself out in the actual history of the Roman conquests. As Rome advanced north into Europe and around the Mediterranean basin it conquered the great forestlands of the ancient world. The forest had served to protect the individual societies and cultures; it was an obstacle to the homogenization of culture—the forest was an "asylum of cultural independence."[70] Thus, the Romans had to tame this mass of forests as they made their empire; and they accomplished the task by destroying forests, by building a complex system of roads, and by installing a common architectural style on their colonies.[71]

Independent Nature Denied

Throughout western history, the taming of the forest is a necessity for the establishment of civilization. Harrison's exegesis on Dante's *Divine Comedy* is particularly apt in this regard.[72] He compares the two forests that Dante as the narrator enters, in the opening lines of the *Inferno* and in the later books of the *Purgatorio*. The first forest is savage and dark, and he loses his way. But the second forest is an earthly paradise, having been denatured, "deprived of its dangers, its savagery, ... its wildlife."[73] Dante is free to wander through this forest and learn its ways. Beatrice even tells him that he will be, for a short time, a forester, and that is how he will become a citizen of Christ's kingdom. For Harrison the point is clear: "to say that the human will has been redeemed means that it has triumphed over nature, mastered its wilderness."[74] Each individual has to control the savage forest within his soul. But the wild untamed forest is also an actual physical reality:

> The forest has ceased to be a wilderness and has become a municipal park under the jurisdiction of the City of God. In Christianity's vision of redemption, the entire earth and all of it nature become precisely such a park, or artificial garden.[75]

Thus Dante's allegory of the will and redemption of the human soul is also a literal description of the progress of society as it seeks to control, re-make, and dominate the natural world. "The earth as a whole must become the legitimate inheritance of humankind."[76]

The domination of physical nature has its source in the control of nature in the realm of science and mathematics—thus recalling a theme of Kidner that nature must first be intellectually mastered before it can be physically destroyed. Harrison uses a discussion of Descartes and his development of a new method for acquiring knowledge to drive home this point.[77] In the opening sections of *Discourse on Method* (1637) Descartes makes explicit use of an analogy with a forest. Those travelers who find themselves lost in a forest ought not to wander this way and that, but instead should attempt to walk a straight path or line, until they can escape and arrive where they are better off. Descartes' new method is an attempt to develop that straight line of clear reasoning, and for Harrison the forest represents for Descartes "all that goes by the name of tradition ... the accumulated falsehoods, unfounded beliefs, and misguided assumptions of the past."[78] Descartes wants to remove the forest in the human mind and replace it with a desert—a vacant plain with no obstructions where the straight lines of geometry can provide order.

Chapter Four

According to Harrison, Descartes' method of reasoning is based on "the mind's abstraction from history," for only pure mathematical reasoning, and the technological power that can be derived from it, constitute the true method for humanity.[79] In this way, as Descartes writes, we "make ourselves the masters and possessors of nature."[80]

All of this exegesis is, of course, metaphor and analogy. The application of the mathematical reasoning of Cartesianism to the actual forests of Europe is another story, but one which can be seen in the birth of the new science of forestry. Here Germany enters our discussion once again, for the use of "forest mathematics" to manage and control the production of timber began there in the latter half of the eighteenth century.[81] The volume of wood production became the guiding principle of forest management. This required the modification of natural forests that were comprised of diverse species and trees in different stages of growth. Instead, new monocultural forests with uniform species and set and regulated plantings were established. These forests produced the maximum yield of timber; they were "an ideal forest whose random and natural variables were reduced to a minimum."[82] They represent the victory of Cartesian thought, for they answer the problem Descartes set in the opening pages of the *Discourse*. As Harrison wryly concludes: "How do you walk in a straight line through the forest? To begin with you plant your trees in rectilinear rows."[83] The domination of the natural forest through abstract reason is complete.

With Descartes we have an appropriate figure to return to our guiding motif, Anne Frank's tree. In 1635 Descartes lived at Westermarkt 6, located around the corner and within a few hundred yards of the Anne Frank house at 263 Prinsengracht. If you are a tourist to Amsterdam nowadays awaiting entry to the Anne Frank House museum exhibit, you stand on a line that snakes along a side block off the Prinsengracht and along the north side of the Westerkerk church. Descartes' former residence is just on the other side of the church block. Few visitors to the Anne Frank House bother to check out the home of the father of modern philosophy. They would be hard pressed to see a connection between the birth of rationalism and the teenaged victim of the Holocaust. But we can now begin to see a connection. The attempt to create an abstract nature totally devoid of sensuous experience, fully explicated by precise rules of mathematics, is a form of domination. It posits a nature that is

Independent Nature Denied

merely a social construction of the human mind; it denies the independent existence of a free and autonomous nature that lies external to the realm of human life. But in Anne Frank's tree we can affirm the resistance to this domination of nature and humanity.

So let us again return to this tree that once grew behind the Anne Frank house, and to the problem I raised at the beginning of this chapter: to be precise, the Anne Frank tree is not a natural tree, for we may assume it was planted and cultivated by human beings in a definite urban environment. Yet as I noted above, it still can serve as a powerful metaphor for the importance of the continued existence of a nature free of human domination. For Anne, nature had no substitute, and she meant by nature the actual physical entities that she could experience—the tree, the birds, the sky—not the nature re-made in the image of human rationality, not the nature that is a construction of human thought and language. Does it matter that this tree, which Anne and I are using as the guiding metaphor in our resistance to the domination of humanity and nature, is not an actual natural tree?

I think not. Consider Harrison's discussion of the idea of nature in the work of Jean-Jacques Rousseau.[84] Harrison focuses on two seemingly contradictory writings of Rousseau, the *Discourse on the Origin and Basis of Inequality Among Men* (1755) and the *Project for the Constitution of Corsica* (1765). In the earlier work we find the well-known view of Rousseau of the importance of "natural man" free of the distortions and corruptions of civilized society. In the *Discourse* Rousseau argues that humans must try to re-connect with the sensuous being of nature guided by the intuition that "natural man" could wander "through the great primeval forests of the earth, living a simple, innocent, and most importantly, *happy* life."[85] But in the *Project for Corsica*, Rousseau takes a strictly utilitarian line, arguing that the resources of nature must be utilized for human good. He advocates a position that is overtly economic in its evaluation of the benefits of the Corsican forests. The trees should be exploited or sold for the good of the Corsican economy. This conclusion appears to be a direct contradiction to Rousseau's more famous position on the significance of undeveloped nature.

How do we resolve the contradiction? Harrison uses a passage from Rousseau's *Confessions* in which he recounts his frequent visits to the Bois de Boulogne, a wooded park on the outskirts of Paris.[86]

Chapter Four

Wandering through the parkland, Rousseau is able to recall the sensuous experiences of an earlier trip to the forest of Saint-Germain in the French countryside. It was in Saint-Germain that he had the vision of a more primitive time and a more natural environment where he could "strip man's nature naked" and see the source of human misery and corruption. The importance of this forest for Rousseau is that it provides a communion with a more primitive nature uncorrupted by human society. It is the direct experience of the Saint-Germain forest that distinguishes it from the forests of Corsica, of which Rousseau had no direct experience. For Harrison, the difference is "the difference between finding oneself inside or outside the forest."[87] From the inside one feels the intuitions of the primeval world; from the outside one just sees the raw material for human exploitation.

Yet there is a profound irony here, as Harrison notes, and it is this irony that will answer our problem of the urban nature of Anne Frank's tree. Rousseau decried the inauthenticity of human life in the civilized world, for the human-built world was a source of alienation from a person's true natural self. Yet Rousseau has this insight, not in the actual wild forests of the undeveloped world, but in the rather tame forest of Saint-Germain and even the woods in the city park of the Bois de Boulogne. This is the blessed forest of Dante, one purified of its savagery, the symbol of the human control of the wildness in humanity and nature! For Harrison this means that

> The ancient state of nature which he [Rousseau] envisions through introspective intuition need not be real or demonstrable. Indeed, Rousseau can even affirm that perhaps the state of nature as he imagines it never truly existed. Yet Rousseau needs the idea or image of that state to denounce his fellow men and their progressive ambitions.[88]

I endorse this conclusion and in turn it apply it to the analysis of the meaning of the Anne Frank tree. The fact that Anne's tree is not really natural is not important. What is important is the use of the idea of the tree as a symbol of resistance and liberation. Although it is a domesticated tree in an urban landscape, for Anne it serves as the representative of a free and autonomous natural world.

Am I myself being seduced by the power of metaphor? Perhaps I am guilty of my own form of epistemological domination, in which I project a personal meaning onto this tree and onto nature. Of course:

Independent Nature Denied

I cannot escape the boundaries of human language and discourse. I cannot escape entirely the social construction of reality. I am using this Amsterdam tree, as did Anne, as a symbol of a nature independent of human power, control, and evil. But I use the metaphor to recognize the independent existence of a nature that is prior and external to the realms of human domination, and it is this nature that must be preserved as a limitation on human power and control.

Notes to Chapter Four

1. McKibben 1989; see Vogel 2002.
2. Vogel 2002, 29–30.
3. Vogel 2002, 31–32.
4. Vogel 2002, 32.
5. Vogel 2002, 33.
6. Vogel 2002, 31.
7. Peterson 1999, 341.
8. Peterson 1999, 353.
9. Peterson 1999, 346.
10. Peterson 1999, 351.
11. Peterson 1999, 356.
12. Smith 1999.
13. Smith 1999, 364–68.
14. Smith 1999, 365.
15. Smith 1999, 366–67, original emphasis.
16. Smith 1999, 373, original emphasis.
17. Ibid.
18. Smith 1999, 374.
19. Ibid.
20. Smith 1999, 375.
21. Smith 1999, 374.
22. Smith 1999, 375, original emphasis.
23. Ibid.
24. Ibid.
25. Smith 1999, 375–76, original emphasis.
26. Schama 1995.
27. Schama 1995, 61.
28. Schama 1995, 7.
29. Schama 1995, 95.
30. Schama 1995, 102.
31. Schama 1995, 103.
32. Schama 1995, 201.
33. Schama 1995, 95.
34. Schama 1995, 154.

Chapter Four

35. Schama 1995, 53.
36. Schama 1995, 201.
37. Schama 1995, 29.
38. Schama 1995, 36.
39. See Katz 1996a.
40. Schama 1995, 36.
41. Kidner 2000, 343.
42. Kidner 2000, 348.
43. Kidner 2000, 345.
44. Kidner 2000, 346.
45. Kidner 2000, 354.
46. Kidner 2000, 355.
47. Soper 1995, 15.
48. Soper 1995, 38.
49. Keeling 2008, 511.
50. Keeling 2008, 510.
51. Evernden 1992, 22–23.
52. Evernden 1992, 23.
53. Evernden 1992, 50.
54. Evernden 1992, 56.
55. Evernden 1992, 60.
56. Ibid.
57. Evernden 1992, 93.
58. Evernden 1992, 94.
59. Evernden 1992, 95–96.
60. Evernden 1992, 110.
61. Ibid., original emphasis.
62. Evernden 1992, 121.
63. Evernden 1992, 116.
64. Evernden 1992, 120.
65. Harrison 1992.
66. Harrison 1992, ix.
67. Harrison 1992, xi.
68. Harrison 1992, 49.
69. Harrison 1992, 38-40. Note that Harrison's view of the Dionysian/Socratic conflict derives from Friedrich Neitzsche's *The Birth of Tragedy*. Harrison only mentions Plato's *Symposium* in his discussion; I also find relevant the passage in *Phaedrus* (230d) where Socrates claims that he cannot learn anything from trees and open country, but only from men in the city.
70. Harrison 1992, 51.
71. Ibid.
72. Harrison 1992, 81–87.
73. Harrison 1992, 85.
74. Harrison 1992, 86.
75. Harrison 1992, 85–86.
76. Harrison 1992, 87.

Independent Nature Denied

77. Harrison 1992, 109–113.
78. Harrison 1992, 111.
79. Harrison 1992, 112.
80. Descartes, quoted in Harrison 1992, 113.
81. Harrison 1992, 122.
82. Ibid.
83. Harrison 1992, 123.
84. Harrison 1992, 125–32.
85. Harrison 1992, 128, original emphasis.
86. Harrison 1992, 129–30.
87. Harrison 1992, 130.
88. Harrison 1992, 132.

~ Chapter Five ~

THE DARK SIDE OF AUTHENTICITY: NATIVISM, RESTORATION, AND GENOCIDE

I.

Recall these words from the diary of Anne Frank: "I hear the approaching thunder that, one day, will destroy us, too, I feel the suffering of millions. And yet, when I look up at the sky, I somehow feel that everything will change for the better, that this cruelty too will end, that peace and tranquility will return once more."[1] We have been using the Anne Frank tree as a symbol of her resistance to the domination and oppression that engulfed her life and the world around her. As such, I have claimed, it can serve as a symbol for us to resist the forces of domination regarding environmental policy, as in the practice of ecological restoration. It is now time to look more closely at the nature of the evil that confronted Anne and the "suffering millions" with which she empathized and identified. What was the "approaching thunder?" Does it reveal a fundamental connection between the domination of nature and the oppression and genocide of humanity? I believe that it does, and that the language, rhetoric, and arguments in contemporary debates over restoration policy clearly exhibit the connection.

In this chapter and the next, we turn to the ethical ramifications and the policy alternatives that can be derived from the more ontological and epistemological topics of the preceding chapters. And we examine what can be considered to be the "dark side" of the positive concept of authenticity that I developed in previous chapters. "Authenticity" can be mis-used as a guiding principle of ethics and policy, and it can lead to disastrous results for both humans and nature. Yet one cannot fake authenticity. To impose an artificial conception of authenticity through force is merely another form of domination. The result, as we will see explicitly in Nazi policies, is eliminationism and genocide.

I will begin this discussion by returning to the topic of the meaning and merits of the practice of ecological restoration. Should the debate over ecological restoration invoke a comparison to Nazism? Is the

The Dark Side of Authenticity

project of the restoration of natural environments somehow connected to the Nazi ideals of "blood and soil," the desire for a pure landscape or home for a native and natural race of humans? Raising this question is extremely provocative, for any comparison to Nazism, in almost any context, tends to produce strong visceral responses. Yet there appear to be enough substantive connections to warrant a critical investigation. There exists a metaphorical comparison of Nazi eliminationist policies regarding specific human populations to the eradication of invasive and non-native species in ecological restorations. Moreover, there are substantive environmental policies of the Nazi regime that appear to be similar to the goals and methodology of contemporary restoration practice. But my argument in this book is that there is also a more fundamental connection: the idea of the domination of the natural world. It is this more fundamental similarity that is most often overlooked in the use of the Nazi comparison to ecological restoration. The idea of domination is the key to understanding both the process of ecological restoration and its real connection to Nazism. The issue here is not so much the alleged purity or authenticity of restored ecological landscapes but rather the continual human project of the management and control of the natural world. The point is not to equate the nativist tendencies of some ecological restoration practices with Nazism, but rather to show that these nativist tendencies are yet one more example of the human domination of the natural world.

A good starting point is a brief editorial written by William Jordan III in response to a *New York Times Magazine* article by popular food and garden writer Michael Pollan.[2] In May 1994 Pollan published a piece titled "Against Nativism," a gentle criticism of what he labeled the "natural garden" movement in the United States. Pollan astutely argued that the natural garden movement was based on an aesthetic ideal of wilderness, for it "outlaws any human artifice in [the] design" of the garden and "grants citizenship exclusively to native plants" for the purpose of creating a garden that "resembles as closely as possible the 'presettlement' American landscape of its particular region." Although this analysis is mild, Pollan went on to compare the rising "intolerance toward foreign species" that is part of the natural garden movement with a similar movement in Germany in the 1930s. There, a so-called "blood-and-soil-rooted" garden sought "to give the German people its characteristic garden and to help guard it from unwholesome alien influences" which might weaken the "Nordic races."[3] Pollan explicitly rejected the idea that he was implying that the natural garden movement in the United States was a "crypto-Fascist" or Nazi movement. He claimed that he was merely demonstrating that the natural garden movement in America had historical precedents (as did the movement in 1930s Germany), but that more importantly there is a danger of "ideology in the garden masquerading as science." This, again, seems a rather mild criticism and a point well taken in a variety of contexts. Any rational thinker should be aware that ideology might affect scientific conclusions and debates.

Nevertheless, Jordan, who many consider the leading proponent of the policy of ecological restoration, responded to the Pollan essay as if restoration were being equated with Nazism. The over-reaction is instructive. Jordan's response,[4] I daresay, is muddled. He first admits to being repeatedly surprised by the comparison with Nazism, "that ecological restoration is a form of nativism—the ecological version of the sort of racist policies espoused by the Nazis or the Ku Klux Klan." Jordan summarizes the attack against restorationists: "Like the Nazis and the Klan, restorationists espouse the exclusion and removal of immigrants, and even a program to ensure the genetic purity of stock in order to protect the integrity of the native, the true-born, the Blut und Boden [blood and soil]." Although he believes that this is a mischaracterization of the restorationist position, he goes on to justify these exclusionary and eliminationist policies by the goal of restoration: "on a purely ecological level ... measures to exclude, eliminate, or control certain exotic species ... [are] necessary if we are to hang onto classic ecosystems." So why is he surprised by the suggestion that ecological restoration is a form of Nazi-like nativism?

Jordan continues in this vein, not denying but rather defending the nativist tendencies of restoration, for the goal of these operations is *"to protect the oppressed and threatened group from extinction"* (original emphasis). He even attempts to subvert the Holocaust comparison by claiming that restoration "is more like the creation of modern Israel" in that it is seeking to provide grounds for the survival of a threatened population. He continues to play with various metaphors of gardening and various social perspectives, claiming that Pollan's ideal garden that

The Dark Side of Authenticity

mixes native and indigenous species is rather like the concept of a social melting pot, which Jordan thinks is racist. He ends his editorial by claiming that we need an "ethic of discrimination" so that not everything is thrown out—which I guess means that we need to know what good species to preserve and what bad ones we can eliminate. He also assumes that Pollan will agree with his conclusion: "I suspect it's not so much restoration and the natural garden he objects to as the tone of voice in which these matters are sometimes discussed."

Actually Pollan never even mentions restoration—he only talks about gardening. So Jordan's muddled response that both denies and argues for the similarities to nativist discrimination is a reaction to a criticism that is never lodged against restoration. Jordan "doth protest too much." The Nazi comparison clearly cuts too close to home for restorationists to be comfortable. We need to examine the substance of this comparison to understand the source of this discomfort.

II.

The central idea under discussion is the control and elimination of non-native (exotic or alien) species from particular landscapes and ecosystems. There are both scientific and cultural (i.e., ethical and sociological) issues involved, and some problems that blend scientific and cultural questions. From the scientific perspective, there are problems defining and identifying native and non-native species, problems in understanding anthropogenic changes in ecosystem populations, and problems in understanding the role of species migrations.[5] There are problems regarding the potential harm that non-native species cause, involving biodiversity, ecological health, aesthetics, and animal welfare. The control and elimination of non-native species also raises questions about the meaning of naturalness and economic well-being.[6] These issues of potential harm, meaning, and economics can be seen as falling within an orientation that is both scientific and cultural. Finally, from the perspective of ethics, sociology, and culture there are a number of problematic issues: the rhetoric of xenophobia and nativism that advocates biological and social purity, the criticism of the global homogenization of regional spaces, and metaphors of invasion, conquest, and imperialism.[7] In what follows, I do not have the time or space to analyze all of

Chapter Five

these concerns. Yet all of these issues, it seems, play into the contested comparison with Nazism.

First, consider the scientific issues. The very identification of species as native or alien/exotic is problematic. Given the obvious fact of species migration throughout history, how do we identify the species in a given area as native or not? As Jonah Peretti notes, we would need a comprehensive natural history of the area[8] but in fact this natural history would necessarily extend backwards in time literally to prehistoric eras. It is clear that we do not have such histories for natural regions that can be considered reliable. And even if we were to have such records, the question arises as to how long a time period a newly migrating species must be in an area until it is considered native? Can a species become naturalized over time within a given area?[9] These issues undermine the very meaning of "native" or "alien" in regard to natural species. Thus Mark Woods and Paul Veatch Moriarty consider five possible (and overlapping) criteria for determining the meaning of an exotic or alien or non-native species: human introduction, evolutionary history, historical range, degradation to the ecosystem, and membership in a community. Each of these criteria presents problems if we are to take it as the basic meaning of a non-native species. None of the criteria appears to be either necessary or sufficient to denote an alien species.[10] Woods and Moriarty thus propose that native and exotic be understood as "cluster concepts" that involve a variety of traits or characteristics generally—but not always— associated with the concept.[11] Five of the traits are those listed above—there could be others—and the more of these traits that a species has the "more likely we are to think of it as exotic."[12] A species introduced by humans into an area outside of its historical range that flourished but also caused degradation to the area ecosystem would be considered more alien than a species that simply migrated into a new area on its own without causing significant ecological disturbances. Being native or alien is thus conceived along a spectrum—species can be more or less exotic.

What this analysis demonstrates is that the scientific understanding of native and non-native species is bound up with significant issues of philosophical meaning. It would be naïve to believe that science can resolve all of the issues that arise when considering native and exotic species. The analysis also shows that there are normative issues in the

The Dark Side of Authenticity

determination of scientific meaning, for one of the criteria listed by Woods and Moriarty is the harm (degradation) caused by the species under question. The scientific issues of anthropogenic changes to ecosystems and historical migration patterns of nonhuman species face similar concerns. Peretti argues that the idea that the European colonization of the Americas degraded the natural ecosystems of the "new world" is based on questionable scientific, anthropological, and normative claims about Native American culture and society. It requires a myth of an idealized society of primitives living in harmony with the natural world, a myth not supported by scientific research. The truth is that anthropogenic changes to the natural environment have existed as long as humans have existed, so that ideas about native and human introduced species are fundamentally unclear.[13] In addition, Peretti argues that recent research in conservation biology presents a new model of species migration, depicting this process as less an unusual invasion and more the normal process of nature. The idea that alien species enter closed climax systems and cause disruptions is no longer a dominant view in conservation biology, because the idea that ecosystems exist in some ideal state of the "balance-of-nature" has been challenged.[14] Citing the work of biologist Rob Hengeveld, Peretti notes "species evolve in unstable conditions that promote tolerance to biological invasions and changing species compositions." Thus, "free species migration [is] a central element in preventing species extinctions."[15]

This means that the science of native and non-native species is indeterminate, and that this indeterminacy will affect the normative and policy issues that arise from the consideration of native and non-native species. When we turn to a consideration of values issues we do so without an objective scientific foundation. Nevertheless, the common perception is that exotic or non-native species are generally harmful to ecosystems and natural areas because of the loss of native biodiversity, degradation of ecological health, extinction of native species, reduction in naturalness, and the diminution of various anthropocentric benefits such as aesthetic enjoyment, recreational opportunities, and economic value.[16] The typical narrative involves the invasion of a non-native species that disrupts the harmonious interactions of the species within an ecological system. Perhaps the exotic species has no predators in the new system, and thus has an explosive population growth; it then destroys—by some form of over grazing—native species in the ecosystem.

Chapter Five

A variation of this narrative has the alien species as a more powerful or opportunistic species that simply drives out a similar species that is native to the area. Exotic species are often just seen as "weeds." Notorious examples abound in the literature of environmentalism, from the feral pigs of Hawaii[17] to the mountain goats of Yellowstone national park[18] to purple loosestrife and Japanese honeysuckle[19] to the infamous kudzu of the southeast United States.

Is this commonly held view correct? Is it true, as Ned Hettinger claims, that "it is well-known that the spread of exotic species has caused—and continues to cause—significant environmental degradation, including extinction of native species and massive human influence on natural systems?"[20] Mark Sagoff presents a strong counter-argument, making five distinct claims against the common sense position. First, although it is true that non-native species may change the dynamic processes and the physical material of an ecosystem, this does not mean that the invasive species has harmed the system, unless we develop a precise meaning of environmental harm that is not circular or question-begging. Harm cannot be defined as the changes caused by a non-native species. It is true that ecosystems change, but both native and non-native species are causes of these changes, and scientists are unable "to determine in randomly selected ecosystems if non-native species as a rule cause more harm than native ones."[21] A related point is that many non-native species have positive effects on an environment; Sagoff cites studies involving the honeysuckle and the zebra mussel.[22] Second, without a clear criterion of harm, and without a clear picture of the precise changes caused by an invasive species, we would need to eliminate all non-native species, an impossible task.[23] A third argument concerns biodiversity, a goal sought by most, if not all, environmental policymakers and conservation biologists. The problem, for critics of non-native species, is that the introduction of non-native species may actually increase species richness and biological diversity. The old model of an ecosystem containing a finite number of niches, so that an alien invader necessarily pushes out a native species, is no longer credible. "Many ecologists ... suggest that the number of species that can reach a site may be the principal factor that limits the number that can take hold there."[24] A fourth point concerns extinctions, the most often cited harm engendered by exotic species invasions. Sagoff's review of the literature also rebuts this claim: "available data supporting invasion

The Dark Side of Authenticity

as a cause of extinctions are, in many cases, anecdotal, speculative and based upon limited observation."[25] There are good reasons to doubt the commonly held view of alien caused extinctions. It might be "intuitively plausible" to consider that predators in small isolated ecosystems (such as islands or lakes) could eliminate a native species that is preyed upon, but it is difficult to imagine how exotic plants, in ecosystems that permit extensive migrations and variations, could be the main cause of a local extinction.[26] Moreover, the migration and invasion of non-native species may, overall, help to avoid extinction, since species endangered in one location may be able to move and survive in another ecosystem.[27] Finally, Sagoff argues that the claim that non-native invasive species are harmful is largely tautological, a result of a stipulative definition that exotic species are harmful because they are non-native. There is no scientific basis for this claim; it is "an example of political advocacy parading as empirical science."[28]

Sagoff's argument underscores the point that the real issue here is our normative or cultural responses to non-native species. Non-native, exotic, alien, or invasive species—whatever term that is used—are simply not liked, and so scientific reasons and environmental rhetoric are employed to justify disapproval. But this entire process feels all too similar to irrational prejudices, such as nativism and racism. First, consider the language. As William O'Brien notes, the terminology of the discussion is laden with overly aggressive rhetoric. Using a metaphor of "invasion" to describe the migration of species into new areas raises connotations of military action and so (supposedly) inspires defensive responses against the invader. Metaphors of "immigrants" or "aliens" are also used to cast negative attitudes upon the non-native species, for they appear to be associated with the "disruptive" and "threatening" presence of newly arrived human immigrants.[29] Indeed, O'Brien's point is that the negative rhetoric regarding non-native species trades on the parallels between human and nonhuman migration into new countries or regions. Exotic species are compared to overly sexual and prolific immigrants that degrade stable native communities.[30] The language of nativism is thus clearly embedded in the debate over non-native species.

The substantive content of nativism is also present: xenophobia and the desire for a culturally pure homeland. Remember Jordan's goal for successful restorations, mentioned above: "measures to exclude, eliminate, or control certain exotic species ... [are] necessary if we are to hang onto classic ecosystems."[31] It is difficult, if not impossible, to understand this goal in a way that is not a call for a pure landscape, cleansed of all foreign material. O'Brien, in his critique of restoration based on the presence of nativist elements, merely notes the "ambivalence" of restoration regarding landscape purity: "the restoration concept retains an ambivalence toward the nature-society dichotomy that permits a longing for lost community purity that guides nativism aimed at both humans and nonhumans alike"[32]; but "ambivalence" appears to be too mild a characterization. Although O'Brien admits that it would be difficult to establish a "direct connection," there seems to be a rhetoric of anti-immigration in some environmentalist doctrines.[33] The goal of a pure homeland was also a primary aim of Nazi policy. As Peretti reminds us, "the purism of biological nativism has historically been associated with fascist and apartheid cultures and governments." The Nazis "attempted to purify nation and nature, by eliminating people and biota that were supposedly not native."[34] All of this lends substantive support to the comparison of ecological restoration to human based policies of xenophobia, nativism, and eliminationism. To determine the meaning and importance of this comparison we will have to examine the motivation, underlying purpose, and moral value implicit in the practices denoted by the comparison.

The practice of gardening can be a useful illustrative example; recall that it was a discussion of the "natural garden" movement by Michael Pollan with which we began. In a garden landscape, human actors consciously direct and mold the processes of nature. To what end, we might ask? If we plan a garden that is to remind us of a certain place, we will have to take measures to eliminate any species—or indeed any physical features—that are alien to that place. If the place that the garden is to evoke is our "homeland" then we are going to adopt a nativist attitude to the creation of the garden: we are going to eliminate any foreign, non-native, or invasive species. We may even require the adoption of processes that are distinctly non-ecological, as when we insist on planting and maintaining a green lawn in a dry climate. Isis Brook defends the practice of creating a garden that reminds us of home, even though she sees the dangerous parallels with human nativism and Nazism. She argues that the primary danger of the Nazi vision is the massive re-creation of landscapes to evoke a German homeland, as when the lands to the east

The Dark Side of Authenticity

were to be "Germanized" as they were conquered and made part of the Reich.[35] Brook needs to show that a nativist conception of a garden can be distinct from this Nazi ideological vision. She rests her claim on the visceral, emotional, and nostalgic importance of the plants of our home. Gardens, and the plants that lie within them, are powerful bearers of emotional significance, of the attachment to place. For people who are separated from their homelands in distinct and different environments, the introduction of non-native species that remind them of home would be a reasonable goal, as long as this re-creation of place is limited to isolated gardens. If the nostalgia is used as a justification for the re-creation of entire landscapes, then we have adopted the Nazi vision.[36] And so the argument concerning non-native species can cut both ways. Although it is clear that non-native species can be eliminated to preserve a sense of a pure landscape of home, alien species can also be introduced to create the sense of a pure homeland from some distant place. But in each case, humans must manipulate the environment so as to maintain the purity of the desired homeland garden or landscape. (I will return to the concept of manipulation in section IV, below.)

The desire for purity brings us dangerously close to a fundamental moral principle of Nazism: racial purity, with the concept of race being strictly a biological construct. So let us move beyond the creation of homeland gardens to the debate over the globalization of uniform landscapes. Here the maintenance of purity and the attachment to place is even more apparent. Ned Hettinger defends an idea of local landscape purity. He first argues that the comparison between xenophobia or racism and biological nativism is too broadly drawn, for there are "some versions of both cultural nativism and biological nativism [that] are rational and even praiseworthy."[37] On the cultural side, Hettinger cites the preservation of indigenous peoples and cultures even though this is "a kind of purism." On the biological side, Hettinger supports the need for the preservation of biodiversity by maintaining local native ecosystems.[38] Although on a local level, the introduction of non-native species will increase biodiversity (as we saw that Sagoff argued, above), in the long run (Hettinger writes):

> The widespread movement of exotic species impoverishes global and regional biodiversity by decreasing the diversity between types of ecological assemblages on the planet ... When this is done repeatedly, ... the trend is toward a globali-

Chapter Five

zation of flora and fauna that threatens to homogenize the world's ecological assemblages into one giant mongrel ecology.[39]

The choice of words here, to say the least, is unfortunate. The Nazis and other racists often use the terminology of "mongrel races" to disparage the kinds of people of whom they disapprove. So to a certain extent, Hettinger is trading on this negative racist rhetoric. And his argument echoes the fear that cosmopolitan homogenization will "undermine" human community by "contribut[ing] to the loss of a human sense of place."[40] A cosmopolitan person, Hettinger claims, will feel no particular attachment to any one nation or region, calling the entire world home, and thus "is less likely to understand, care about, or defend local cultural practices or biotic communities."[41] The massive extent of globalization and homogenization of bioregions in the contemporary world requires us, then, to defend native flora and fauna and native ecosystems. Hettinger concludes that we are justified in opposing the introduction of exotic species, even when they migrate under their own power and cause no significant physical damage to their new ecosystems.[42] In this way we will preserve a myriad of distinct places and homes, biodiversity and cultural diversity.

Hettinger attempts to address the critical claims of racism and nativism that seem to resonate in his position. His defense is based on the differing motivations of biological nativists and human racists. The biological nativist has the "commendable desire for local biotic purity" while the racist has the "contemptible desire for human racial purity."[43] But claiming that local biological purity is commendable seems to be the point at issue; Hettinger appears to be making a stipulative claim. He argues that biological nativists are not similar to racists, who fear and dislike other races, because the biological nativist appreciates foreign and exotic species in their natural habitats, and they have good reasons—the defense against biological homogenization—for advocating the exclusion of non-native species from local ecosystems. Yet as O'Brien responds, Hettinger's argument merely "reframes what might otherwise appear to be a reactionary anti-exotics argument and presents it as an argument for justice." But the argument "remains rooted in rigid dichotomies that presume to distinguish the purity of the local from the contaminating influences of the outsider" and thus it "replicates many of the same problematic tendencies of the more common reactionary view of biological

The Dark Side of Authenticity

invasions."[44] Both the language and the substantive conceptual framework of the anti-globalization argument for the exclusion of non-native species reinforce the nativist ideal of a pure regional homeland or ecosystem.

The preliminary conclusion is that policies of restoration and ecological management that have as a goal the elimination of non-native species—similar to so-called natural gardening—cannot escape comparisons with nativist, racist, and even Nazi ideals. Our stance towards non-native, exotic, alien, or invasive species is essentially a philosophical-political-cultural position, for the science regarding these species is either unclear or heavily laden with non-scientific cultural values. Prominent among these values is the desire for a pure physical and cultural homeland, the desire for a recognizable place, free of foreigners and their influence, which we can call home. In the following section, I will examine the history of Nazi environmental policies in order to complete the comparison. Yet I will argue in the concluding section of the chapter that the creation of a pure homeland is not the fundamental issue at stake in an analysis of restoration and management policies. Thus I am not arguing that the nativist practices of ecological restoration are morally equivalent to Nazi policies of elimination and genocide. The real issue is the continual project of the domination of nature.

III.

In July 1935, during the early years of the Third Reich, the Nazi regime instituted a comprehensive law for the protection of nature, known as the Reich Nature Protection Law (RNG). The provisions of the law established a national environmental policy, superseding the previous German system of state-by-state regulations and enforcement. Of equal importance to the national scope of the law were provisions that protected landscapes that were "free nature," or that were aesthetically pleasing, or that were in the interests of animals. Officials in the new national office were empowered to issue decrees to enforce protection; they could create protected nature areas and could seize private land in order to do so. Moreover, one section of the law denied indemnification and compensation for those property owners whose land was expropriated. A further section of the law required all government officials—national, state, or local—to consult with the nature protection office on any project that could alter the "free landscape."[45] As historian Charles Closmann notes,

Chapter Five

"these features of the RNG made Germany more progressive in matters of conservation and landscape planning than other industrialized nations."[46]

The idea that the evil Nazi regime was a progressive leader in the development of environmental protection is obviously disturbing to contemporary environmentalists. But the connection of environmentalism to the Nazi regime has been a matter of some debate among historians of the period. Primarily, there are issues in the extent and the sincerity of the Nazi policy of nature protection. How much of the German conservation movement's alliance with the Nazi regime was just opportunism? Was the adoption of the rhetoric of Nazi ideology by leading conservationists merely a political maneuver? Were environmental considerations—and indeed, the 1935 RNG itself—ignored once militarization became the dominant policy of the regime?[47] Secondarily, there are issues concerning the supposed connection of Nazi environmentalism to the contemporary Green movement, a question first raised by historian Anna Bramwell in two controversial books published in the 1980s. Bramwell focused heavily on Nazi agricultural policy, and particularly the ideas and work of Richard Darré, popularizer of the concept of "blood and soil," to argue that there is a fundamental connection between Nazi ideology and contemporary environmentalism.[48]

In this chapter and book, I resist drawing conclusions about this debate among historians, for it is outside my field of expertise. Yet just the existence of the debate demonstrates that there is at least a superficial compatibility of Nazi ideology and contemporary policies of environmental protection. Whether there is some fundamental convergence of Nazi ideas and environmentalism, or whether there is only—in historian Frank Uekoetter's words—an "enduring fragility of the intellectual bridge"[49] between the two, it is clear that basic Nazi ideas and environmental concerns have a good deal of cross-resonance.

First among these ideas is *Heimat*, or "the love of the regional homeland."[50] Although this idea may have been broadened and become abstract during the Nazi regime, embracing the notion of a national homeland, the original meaning of a connection to specific regional and local landscapes clearly has environmental overtones. *Heimat* is what nature protection is all about: the defense and preservation of local ecosystems and what we now call bioregions. Yet the concept was racialized under Nazi ideology. In discussing the life and work of Alvin

The Dark Side of Authenticity

Seifert, the chief proponent of landscape planning in the Third Reich, historian Thomas Zeller explains that lovers of landscape in Germany believed that there was a "connection between the landscape and the human soul." Moreover, landscape "embodied the values of a specific community whose characteristics were increasingly coming to be seen as based on race."[51] Thus the love of landscape—*Heimat*—is intertwined with beliefs about the proper inhabitants of a region, and foreshadows policies that will exclude those who do not belong. As historian Thomas Lekan notes, the concept of *Heimat* protection "enabled the individual to feel himself [as] a part of the larger community,"[52] but this is a community that is understood through the concept of race. This is one meaning then of "blood and soil," a landscape and racially determined community.

For the Nazis the concept of community embraced the idea of *volksgemeinschaft*, the community of the *volk*, the true and authentic people of a region and nation, bound together in a classless community of social equality. The concepts of *volksgemeinschaft* and landscape protection were fused together inextricably in the Nazi regime. As Closmann explains, Walter Schoenichen, head of the Prussian office that oversaw the protection of natural monuments and one of most important Nazis in the German environmental movement, claimed that the existence of a *volksgemeinschaft* required the development of the *volk*'s "nature-loving soul." And the Nazi regime, unlike the Weimar democracy, would accomplish this "because of their commitment to a *volksgemeinschaft* rooted in the soil of the homeland."[53] As we will see below, the idea of the *volksgemeinschaft* even played a fundamental role in the development of Nazi forest policy.[54] Homeland (*Heimat*) and community (*volksgemeinschaft*) thus form the basis of Nazi ideas about the protection of nature and the development of landscapes, although in the Nazi ideological universe these ideas are inseparable from concepts of race.

A third idea in the German conservation movement was anti-materialism, or anti-consumerism, and a desire to move away from the utilitarian based "wise use" of resources conception of conservation that was prevalent in the United States at that time.[55] Here again we see a common environmentalist theme, but even this idea is used racially by the Nazi regime. Nazi anti-materialism is connected to the *volksgemeinschaft*, the community of social equals. An authentic member of the *volk* would place the interests of the community above personal goals. Individual profit and the acquisition of private goods are seen as social evils. The chief producers of these evils are the Jews with their insatiable desire for gold, money, and economic power; thus under the Nazis even the environmentalist goal of a less materialistic and consumerist society is understood through the category of race.

In sum, the Nazi regime created a linkage among concepts of the *volk*, racism, and conservation as a means of understanding the German landscape as a homeland, or *Heimat*.[56] How do these basic ideas of Nazi environmentalism play out in the policies of alien or non-native species? First consider the idea that "gardening" was a central metaphor of the Nazi regime. As Zeller notes, "garden tending was akin to nation building: desirables were to be cultivated, undesirables weeded out."[57] Of course, this idea of gardening is only a metaphor, and all by itself it does not lead us to any conclusions about the role of exotic and non-native species in the German landscape. Uekoetter, who argues that the Nazi-environmentalist connection is fragile, claims "there was never a uniform opinion on nonnative species among German conservationists of the interwar years."[58] Yet Uekoetter cites Zeller's work on the life of Nazi landscape planner Alvin Seifert as an example of the way that Nazi ideology informed environmental policy. Seifert "became increasingly radical in his attacks on nonnative species" and he fused his scientific objections with Nazi rhetoric because of political expediency.[59] Seifert was a disciple of Willy Lange, a well-known garden architect and designer in Germany in the early twentieth century. Lange based his views on three central ideas: the science of ecology, what he called biological aesthetics, and race. Thus, "the garden [should] be created as part of the landscape in which it is situated ... and the German *volk* had to play a special role." Lange's "nature garden is the racial expression of nature."[60] Seifert popularized Lange's ideas in the 1930s, as he "consciously introduced the criterion of nativeness, of being rooted in the soil, into garden architecture."[61] And as the primary landscape planner for the Nazi regime, Seifert had the opportunity to implement his ideas, when for example, he advised Fritz Todt on the construction of the autobahn—although Todt did not completely follow his advice.[62]

In 1934, the regime passed a law Concerning the Protection of the Racial Purity of Forest Plants, mandating the use of only the best phenotypes for "certified seed production."[63] Although there were sound

The Dark Side of Authenticity

ecological and forestry-related reasons for such a law—the avoidance of unhealthy specimens as the source of tree production—the connection to racist rhetoric and ideology is obvious. Moreover, landscape policy was not just the elimination of undesirable species but also the addition of desirable native species, even those that had become extinct. Seifert again claimed "in the landscape 'nothing foreign' should be taken in 'but nothing native must be left out.'"[64] So Herman Göring, in his position in charge of all of German forestry, sought to re-introduce extinct species, for "a true German wilderness would be home to a number of animals that had long disappeared from the German heartland."[65] Göring's motivation may have had more to do with his plans for hunting rare game, but once again the policy is connected to the ideology of a particular German homeland, filled with natural and native species.

The pervasive role of Nazi ideology in environmental policy during this period may also be seen in the idea of the *Dauerwald*—the eternal forest—as the basis of forestry policy and planning. Throughout the Nazi period in Germany, there is an attempt to create a new kind of *volkisch* forest policy opposed to the traditional German scientific and utilitarian forestry of the nineteenth century. *Dauerwald* was to be focused on the management of the "forest organism"—i.e., the forest ecosystem—rather than on individual trees.[66] Today, we might call this ecological holism. *Dauerwald* fit neatly into the propaganda and ideology of the German state, especially the idea of "co-ordination" or *Gleichschaltung*, for it meant that the forests were to be managed for the productive benefit of the entire national community—the *volksgemeinschaft*—not for the profit of the individual property owner.[67] Moreover, the analogy between the forest and its individual trees and the nation and its individual citizens helped the Nazis "naturalize the idea of the *volksgemeinschaft*, their ideal of a classless, racially pure, and 'eternal' ... national community." The *Dauerwald* was "an 'organic' structure" containing only native species, and it was "a collective and perpetual entity that had no fixed morphology or lifespan,"[68] similar to the Third Reich. Thus combined with the 1934 law concerning the racial purity of forest plants, even forest policy advanced the idea of pure German *Heimat*, a racialized national landscape. The forest landscape in general and a particular kind of forest organization served as an ideological model for the political regime. As one German

Chapter Five

forestry journal noted in 1939: "Ask the trees, they will teach you how to become a National Socialist."[69]

The plans for the Eastern expansion are the culmination of this connection between Nazi ideology and environmental policy. The ideas of nativist landscape planning cannot be avoided here. Because of the conquest of lands to the east from 1939 to 1942, the policy of *Lebensraum*—living space—became a reality. But in the words of Seifert, "the entire landscape must be Germanized." Merely eliminating the influence of the Polish community would not be enough: the landscape would have to be reconstructed "along Nazi ideals."[70] Seifert's chief rival in the field of landscape planning, Heinrich Friedrich Wiepking-Jürgensmann, was chosen by the Reich Commissariat for the Strengthening of German Nationality to oversee the planning.[71] Although a professional rival, Wiepking-Jürgensmann shared Seifert's ideas about nativism in the landscape movement. Wiepking-Jürgensmann advocated "the idea that Germans had a close relationship to their home landscapes, and argued that it was necessary to replicate these homelands in the conquered territories."[72] The work in the eastern territories permitted the complete freedom of the landscape planner to implement his ideals.[73]

The result was the General Plan East under the direction of Konrad Meyer, the Reich Commissariat's chief of planning, but formally led by Heinrich Himmler himself. According to Uekoetter, what was most disturbing about the plan was its professional technical expertise, concentrating on the creation of a new German landscape without any regard for the human destruction the plan would entail. The plan was "state of the art" dealing with water and soil conservation, planting, clean air, and other environmental policies. Yet "the overarching goal … was to make the land suit a purported German national character, 'so that Germanic-German man feels at home, settles down, falls in love with his new *Heimat* and becomes ready to defend it.'"[74] As we saw in Chapter One, architectural historian Robert Jan van Pelt, quoting from a contemporary source, describes a trip through Poland in 1940 by Himmler and his friend Henns Johst. They stand in a Polish field, holding the soil in their hands, and dream of the re-creation of German farms and villages, the replanting of trees, shrubs, and hedgerows, and even the alteration of the climate by increasing dew and the formation of rain clouds.[75] But a cleansing of the non-German population, the

The Dark Side of Authenticity

elimination of all that was non-native to the Germanic character of the new homeland, was a requirement for the re-making of the landscape. The original plan for the East proposed the transfer of thirty-one million inhabitants to Siberia.[76]

Thus the development of a German homeland to the East was "predicated on the use of violence"[77]—although even this terminology seems too mild. The plan assumed that the entire Jewish population of the area, roughly 560,000 people, would be removed, and that the eventual deportation of 3.4 million Poles would also be necessary; in short, the "plan was to create a new home for ethnic Germans by rendering millions of people who lived in the Annexed Eastern Areas homeless."[78] Himmler, of course, infamously declared that "the destruction of thirty million Slavs was a prerequisite for German planning in the east," and so Uekoetter concludes that the "plan's genocidal implications are obvious in retrospect."[79] Whether or not the plan was consciously connected to the Final Solution, however, is a matter of some historical dispute, and even the two historians that I have relied on for much of this account (Uekoetter and Wolschke-Bulmahn) disagree on this point.[80] As a philosopher and not an historian, I take no position on this issue. Yet even without an explicit connection to the death camps, the Germanization of the eastern conquered lands was able to proceed because the landscape planners embraced the Nazi ideology that the Polish and Jewish populations were sub-human. Wiepking-Jürgensmann "emphasized that an absolute freedom over property was the defining characteristic of conditions in the Annexed Eastern Areas,"[81] so that, as architect Walter Wickop explained, the planning could be accomplished "where one does not have to take anything into consideration"[82]—namely human beings, their rights, their property, their lives. The only issue was the use of the land itself.

In sum, the conquest and domination of the lands to the east were meant to be the creation of another *volkisch* environment, a German homeland, racially and ecologically cleansed of all alien elements. The extreme nationalism and racism of Nazi ideology were expressed in environmental and landscape planning policies. Violence, forced evacuations of local populations, murder, and imperialism all served the goal of creating an authentic German landscape. This re-making of the landscape must be considered a *restoration*: the point was to return to an

Chapter Five

authentic German past, purified of all external non-German elements. It is clear then that an emphasis on native species and the elimination of exotics in the contemporary practice of ecological restoration can meaningfully be compared to Nazi policies of exclusion and eliminationism. Yet I want to argue that this connection of nativist landscape planning to Nazi ideology is not the main problem with the policy of ecological restoration. The nativist tendencies of some forms of ecological restoration represent merely one instance of the more general problem with all varieties of ecological restoration. The real issue is the idea of domination, the coercive control of humanity and nature. In the final section of this chapter I turn to this argument.

IV.

Consider the Nazi idea of "blood and soil" as a foundation for Nazi policies of imperialism, environmentalism, landscape planning, and population control. Richard Darré's idea of "blood and soil" was meant to convey the necessary connection between a particular race of people and the land on which they lived and worked—and most importantly for Darré, the German peasant and the agricultural land of the German empire. There is, however, an ideological contradiction in the relationship of "blood and soil" and Nazi environmental and landscape policy. "Blood and soil" is a fundamentally incoherent concept because it seeks to maximize two opposing forces: the genetic roots of a national community (blood) and the formative properties of the national or regional environment (land). In short, is the natural essence of a *volk* the result of an absolute and unique genetic character or the result of "the material conditions of the natural-geographical environment in which the life history of the *volk* had unfolded?"[83] As historian Mark Bassin notes, it is impossible to reconcile the geographical determinism of the concept of the land as formative of a national character with the racialism of the genetic determinism of blood.[84] For the Nazis—although not for us—there is a contradiction because of the primacy of the concept of race (blood) as the ontological determination of value. Race or blood should make a man what he was irrespective of what land was his home. An Aryan would be an Aryan even if he lived in the Arctic. A Jew would be a Jew even if had lived his entire life in the German ancestral landscape. So why then should land, soil, or home be important? This conceptual tension

The Dark Side of Authenticity

between blood and soil continues to play itself out during the decades prior to the Nazi ascendancy to power, and is only resolved in practice by the political will of the Nazi regime as it privileges the concept of race over that of the environment.

The racialist supremacy of blood is maintained by two additional ideas: first, there is the acknowledgement that land or the environment can affect the genetic character of a race, but that this happened only in primordial times, in the prehistoric process of racial formation; once the genetic characteristics of the race are initially formed, the environment can have no effect.[85] Second, the inner spirit of the *volk* is stressed as the main causal agent affecting the land, as Darré stated: focus on "the special way in which a *Volk* itself shapes its relationship to the soil, and in what form it owns and manages the native soil" because the land "is subordinated to the laws of the life forces of the *Volk*."[86]

It is this second idea—the inner spirit of the *volk* or "blood" is the dominant causal agent for the transformation of the land—that is central to my argument. It means that human power, exemplified by the essence of the *volk* community, can control, manipulate, and dominate the natural world. Human beings and human essence is the fundamental causative force in the world, and it is the human mission to transform the world according to its guiding ideology. As Bassin notes, "the anthropogenic domination of the natural world was an essential part of the activist ethos of National Socialism." The German race could use its creative essence to master the natural landscape.[87] This meant that ultimately the natural landscape falls away, becomes converted to a cultural, human landscape. Bassin again: "The *Volk* ... exercise[s] its formative agency, by altering and shaping that natural world in its own image and for its own purposes."[88]

In the preceding section of this chapter, we have seen what these purposes were: the creation of a pure German homeland. The new German *Heimat* would not be a natural landscape, but a cultural landscape that expressed the highest cultural ideals of the German racial community. In reviewing that history of Nazi environmentalism and the expansion to the East, it appeared that the cleansing of the landscape, the establishment of a pure racial population in a pure nativist environment, was the primary evil of this policy. Piers H. G. Stephens has argued that both Nazism and some extreme forms of environmentalism share a common belief in a "dangerous purity" of the natural and social world, dangerous

Chapter Five

because it demands a purity that is "absolute."[89] Clearly this purity is an essential element of Nazi ideology concerning race, and as we have seen, this goal of purity is definitely employed by Nazi landscape planners as they envision a restored Germanized homeland to the East. But was "purity" the primary determinant of the evil normative character of Nazi environmental and landscape policies?

Pure blood, the racial essence of the *volk*, was indeed privileged in the Nazi ideological worldview. Yet the *volk* was meant to change and shape the landscape to meet the needs of the *volksgemeinschaft*, the racially and socially united community. What is crucial is the power of the *volk*, the power of the human race. Soil, so to speak, is only important as an expression of the blood of the *volk*, the product of the will of the racial community to shape and structure its world. Soil, land, and nature are not central to an ideology of *volkisch* domination: blood, race, and humanity are central. A commentator on the Nazi plan for the East, Lutz Heck, wrote in 1942: "for the first time in history, a nation is undertaking the modeling of a landscape in a conscious way."[90] Whether or not this historical claim is true is not important. The key point, as Bassin noted, is that the Nazi plan involved the *conscious* human control and domination of the natural landscape, the transformation of the natural landscape into a human or cultural landscape. The result of this conscious control was, of course, horribly evil: the attempt to establish a racially and environmentally pure German landscape—yet in my view it was the process itself, the attempt to dominate and control the natural and human world, that was the primary evil. The real problem is that the world would be transformed according to a human ideal, and any human ideal can fall victim to a vicious ideology or worldview. So even if the goal of the transformation were the production of human good, it would still have been an example of human domination. And even if the goal of the transformation were the repair of human-induced harm to the natural ecosystem, the so-called "good for nature itself," this would still be a goal determined by human ideals of what nature should be. As we saw in section II of this chapter, the entire idea of a native natural system, cleansed of human-introduced "exotic" species, is problematic—it is a human determination. Moreover, as we saw in Chapter Three, the process of ecological restoration itself is an expression of domination. Restoration results in a humanized world that has been manipulated to

The Dark Side of Authenticity

fulfill a conscious human ideal. It is this conscious transformation of the environment that is echoed in the Nazi plan to re-make the East into a renewed authentic German landscape.

Consider one final case from the history of Nazi environmental planning: the regulation of the Ems, a major river in northwest Germany.[91] As we will see, this is not a case of ecological restoration, but rather a case in which there is the explicit and conscious control of the natural landscape. I introduce it here to demonstrate that manipulation and control of natural processes—the domination of nature—is the central problem in the human relationship to the natural world. Beginning in the 1920s, plans were developed for controlling the amount of water in the watershed area of the Ems as a means of preventing devastating floods after heavy rainfall that were damaging to the farmland in the region. Hydrological engineers were in charge of the various plans, and they "envisioned a waterway that bore more resemblance to a channel than to the wild, scenic river that the Ems still was."[92] Uekoetter's narrative of this case history focuses on the ways in which the German conservation community attempted to use the Nazi rhetoric of the overall good for the *volkisch* community as a means of preserving the aesthetic and recreational character of the scenic river, but to no avail. Publicly, Nazi officials expressed a "commitment to conservation ... but the actual work along the river banks revealed little in the way of an environmental ethic," for the river was re-positioned and re-dug so as to run in straight lines with the appearance of a canal.[93] Thus one can argue, as Uekoetter does, that the connection between environmentalism and Nazism was not especially strong, for environmentalist goals were quickly sacrificed for other economic and political ends. For me, however, this case is also important because it demonstrates what happens with the imposition of human intentions onto a natural landscape. We do not need to look at an extreme case of the Germanization and purification of the conquered eastern territories to see the moral problem with the conscious human transformation of the environment. Even in a relatively benign case such as the Ems river project, with the goal of saving agricultural land from damaging floods, we see the danger of the humanization of the natural landscape: a beautiful scenic river is transformed into a utilitarian and artificial canal. Adopting the policy of the human control of the natural world, we will create a world in which no landscape is wild and free,

Chapter Five

but all is a product of human conscious design. We will have gardens, farms, and zoos—all humanized landscapes—but no meadows, forests, or grasslands filled with wild beasts.

There is a clear lesson here for the contemporary policy of ecological restoration: the planning of landscapes involves the control and domination of the natural and human world. This control and domination must follow some set of ideological principles, for it is these ideological principles that will determine the goal or plan of the restored landscape. These principles, of course, may be good, or evil, or even neutral. In the history of Nazi environmental planning, we see an extreme example of a guiding ideology that is evil, and there is no reason that ecological restoration today needs to follow the *volkisch* agenda of Nazism. Yet even if ecological restoration avoids the nativist rhetoric and policies that seem to be implied by the elimination of exotics, there remains the more fundamental issue of the humanization of the natural world. The ideology of restoration is not the advocacy of the "purity" of native and natural landscapes, but rather the supremacy and power of humanity, the belief that the entire world can be altered and managed according to human ideals. The removal of species that humans do not want in a landscape, or the re-introduction of species that humans do want, is not a return to a landscape that exhibits less human manipulation, as some have argued[94]—it is simply another form of control. The attempt to correct human-induced harm to the environment is still an example of human control, for it is we humans who determine which species or landscape structures need to be eliminated or preserved. The Nazis, after all, were attempting to correct the human-induced harm to the Germanic landscape that was caused by the influence of non-Aryan Slavic and Jewish populations. Therein lies the similarity of ecological restoration to Nazism: the belief in human power to control the world. Ecological restoration is a policy that seeks the conscious transformation of the natural world into a human and culturally determined landscape. Nazi environmental policy in the East was also the attempt to create a culturally determined landscape. The true connection between the practice of ecological restoration and the Nazi worldview is the policy of human domination—the domination of both the human and the natural world.

And so once again we return to the meaning of Anne Frank's tree. As a symbol of a free autonomous nature standing in opposition

The Dark Side of Authenticity

to the domination of humanity it appears particularly appropriate to an examination of the connections between the practice of ecological restoration and the environmental policy that partially underlies the Nazi genocide in Eastern Europe. Anne's tree is a symbol of that authentic nature that exists outside the conscious manipulation and control of humanity while at the same time it is a symbol of the resistance to the evils of Nazism. These two symbols are inextricably linked in meaning. The biological racism that was the heart of Nazi ideology led to an attempt to impose a bogus authenticity to the German homeland. The cleansing of all foreign elements—nonhuman natural entities and human populations—from the landscapes to the east denied the historical development of both nature and humanity in those regions. It was an attempt at "authenticity" and "purity" that was imposed by force, and thus was a form of domination. It is similar to those aspects of restoration policy that attempt to create pure and authentic classic landscapes; in reality, the restoration policy creates only artifacts that resemble the authenticity of the land and its inhabitants. Real authenticity, as we have seen, is exhibited by the free and autonomous development of nature and humanity. It cannot be imposed from without, by the domination of humanity. Anne's tree—as a symbol of authentic nature—reminds us to resist the forces of domination, and in her own words, "to feel humble and ready to face every blow with courage!"[95]

Notes to Chapter Five

1. Frank 1991, 333: 15 July 1944.
2. Jordan 1994 and Pollan 1994.
3. Pollan cites Groening and Wolschke-Bulmahn 1992 for his information on the German natural garden movement.
4. Jordan 1994, 113–14.
5. Peretti 1998; Woods and Moriarty 2001.
6. Woods and Moriarty 2001; Hettinger 2001; Sagoff 2005.
7. O'Brien 2006; Hettinger 2001; Peretti 1998.
8. Peretti 1998, 185.
9. Peretti 1998; Hettinger 2001.
10. Woods and Moriarty 2001, 164–73.
11. Woods and Moriarty 2001, 174–76.
12. Woods and Moriarty 2001, 175.
13. Peretti 1998, 186.
14. Peretti 1998, 186–87.
15. Peretti 1998, 187.

Chapter Five

16. Woods and Moriarty 2001, 181.
17. Woods and Moriarty 2001, 180.
18. Hettinger 2001, 194.
19. Sagoff 2005, 220.
20. Hettinger 2001, 194.
21. Sagoff 2005, 219–220.
22. Sagoff 2005, 221.
23. Sagoff 2005, 223.
24. Sagoff 2005, 225.
25. Sagoff 2005, 226, citing J. Gurevitch and D. K. Padilla.
26. Ibid.
27. Sagoff 2005, 227.
28. Sagoff 2005, 228.
29. O'Brien 2006, 71.
30. O'Brien 2006, 66–69.
31. Jordan 1994, 113.
32. O'Brien 2006, 65.
33. Ibid.
34. Peretti 1998, 188.
35. Brook 2003, 230.
36. Brook 2003, 231–33.
37. Hettinger 2001, 215.
38. Hettinger 2001, 215–16.
39. Hettinger 2001, 216.
40. Hettinger 2001, 217.
41. Hettinger 2001, 218.
42. Hettinger 2001, 219.
43. Ibid.
44. O'Brien 2006, 73. See Brennan 2006 for an attempt to move beyond these dichotomies and bring together the strengths of both extreme positions.
45. Closmann 2005, 20–21; Uekoetter 2006, 65.
46. Closmann 2005, 21.
47. Lekan 2005; Zeller 2005; Uekoetter 2006.
48. Bramwell 1985; Bramwell 1989.
49. Uekoetter 2006, 36.
50. Uekoetter 2006, 20.
51. Zeller 2005, 150.
52. Lekan 2005, 81.
53. Closmann 2005, 27–28.
54. Imort 2005, 52.
55. Uekoetter 2006, 21; Closmann 2005, 24.
56. Brüggemeier, Cioc, and Zeller 2005, 8.
57. Zeller 2005, 147.
58. Uekoetter 2006, 14.
59. Uekoetter 2006, 79.
60. Groening and Wolschke-Bulmahn 1992, 120.

61. Groening and Wolschke-Bulmahn 1992, 121.
62. Zeller 2005, 151–54.
63. Imort 2005, 61.
64. Groening and Wolschke-Bulmahn 1992, 121–22, citing a 1937 paper by Seifert.
65. Uekoetter 2006, 103.
66. Imort 2005, 47.
67. Imort 2005, 51.
68. Imort 2005, 52.
69. Imort 2005, 54.
70. Wolschke-Bulmahn 2005, 245.
71. Uekoetter 2006, 157.
72. Wolschke-Bulmahn 2005, 246.
73. Uekoetter 2006, 157.
74. Uekoetter 2006, 158.
75. van Pelt 1994, 101–103.
76. Uekoetter 2006, 158.
77. Wolschke-Bulmahn 2005, 247.
78. Wolschke-Bulmahn 2005, 248.
79. Uekoetter 2006, 158–59.
80. Uekoetter 2006 takes the negative position; Wolschke-Bulmahn 2005 takes the affirmative.
81. Wolschke-Bulmahn 2005, 250.
82. Wolschke-Bulmahn 2005, 252.
83. Bassin 2005, 207–08.
84. Bassin 2005, 209.
85. Bassin 2005, 216.
86. Bassin 2005, 224.
87. Bassin 2005, 216.
88. Bassin 2005, 229.
89. Stephens 2000, 272.
90. Uekoetter 2006, 160.
91. Uekoetter 2006, 109–25.
92. Uekoetter 2006, 112.
93. Uekoetter 2006, 120.
94. See Sylvan 1994, Light 2000, Higgs 2003; and the discussion in Chapter Three.
95. Frank 1991, 319: 13 June 1944.

~ Chapter Six ~

ETHICAL CODA: THE NAZI ENGINEERS AND TECHNOLOGICAL ETHICS IN HELL

I.

How can we begin to use the courage of resistance bequeathed to us by the symbol of the Anne Frank tree? To resist the human domination of nature is not a clearly defined action. What does one do? To resist the oppression and domination of humanity might, perhaps, be clearer. In this chapter, then, I address the case history of the Nazi engineers and other technological professionals who worked within and for the Nazi system of domination, oppression, and genocide. Although this case history has no explicit connection to the instances of the domination of nature that we have examined in this book—ecological restoration or the Nazi re-construction of the European landscape—it does connect to the issue of the ethics of technology that was initially discussed in Chapter Two. The fundamental error in assessing the human application of technology is to think of the technological instruments and artifacts that we use as value-neutral. We will be able to observe this error quite clearly in the actions and beliefs of the Nazi engineers; but we must also realize that the same kind of thinking can be applied to the human domination of the natural world, and the application of science and technology to make nature into a system of artifacts. To understand the ethical mistakes of the Nazi engineers then can be a lesson for resisting all the forces of technological domination, applied to both humanity and nature.

Let's begin with this fact: engineers, architects, and other technological professionals designed the genocidal death machines of the Third Reich. The death camp operations were highly efficient, so these technological professionals knew what they were doing: they were, so to speak, good engineers. As an educator at a technological university, I need to explain to my students—future engineers, industrial managers,

Ethical Coda

and architects—the motivations and ethical reasoning of the technological professionals of the Third Reich. I need to educate my students in the ethical practices of this hellish regime so that they can avoid the kind of decisions and ethical justifications used by the Nazi engineers. In their own professional lives, my former students should not only be good engineers in a technical sense, but good engineers in a moral sense.

In this chapter, I examine several arguments about the ethical judgments of professionals in Nazi Germany, and attempt a synthesis that can provide a lesson for contemporary engineers and other technological professionals. How does an engineer avoid the error of the Nazi engineers in their embrace of an evil ideology underlying their technological creations? *How does an engineer know that the values he embodies though his technological products are good values that will lead to a better world?* This last question, I believe, is the fundamental issue for the understanding of engineering ethics. It is a question, perhaps, that is unanswerable.

One terminological clarification before I begin the argument. In this chapter I use the broad term "technological professionals" to refer to those professionals who design, create, and use technologies, technological products, and technological artifacts. Included in this class of professionals are engineers, architects, and specific kinds of industrial managers. These professions are similar to the professions of medicine and law, in that they not only employ specific technical expertise but they also provide a significant social role and purpose. I do not include various types of technicians who mainly deal with the operation, installation, and repair of technological products.

II.

First let us consider the Nazi doctors. Although physicians are not technological professionals as I have stipulated them above, they are professionals who use science and technology to perform the specific tasks of their occupation. An analysis of the Nazi doctors provides a useful starting point for an understanding of the possibility of moral evil in technology. In his groundbreaking account of Nazi physicians Robert J. Lifton proposed the concept of "doubling" as a psychological explanation for the behavior of the medical professionals associated with the worst aspects of the Third Reich: the T4 euthanasia program, the coercive and inhuman medical experiments, and the operations of the

Chapter Six

death camps.[1] Lifton asks the question, how could well-educated professional people act in such horrific and immoral ways, and his answer is that they created for themselves two selves, two personalities, each of which would control different aspects of their lives. For Lifton, what he calls "doubling" is "the division of the self into two functioning wholes, so that each part-self acts as an entire self." Thus "an Auschwitz doctor could, through doubling, not only kill and contribute to killing but organize silently, on behalf of that evil project, an entire self-structure … encompassing virtually all aspects of his behavior."[2]

Lifton differentiates doubling from the more common notions of "split" personalities and psychic numbing, by which the Nazi doctors could suppress their feelings in relation to murder. The doubled personalities are holistic, in that they are full functioning selves adapting for a year or more in an environment that is solely organized around killing.[3] In short, nothing is being suppressed by the doubling agent; rather all feelings and beliefs are being transferred to another fully functioning self. Thus, the Nazi doctors were able to avoid guilt "not by the elimination of conscience but by what can be called the *transfer of conscience*. The requirements of conscience were transferred to the Auschwitz self, which placed it within its own criteria for good"[4]—in this case, the basis of proper behavior within the killing system of the death camp: e.g., "duty, loyalty to group, 'improving' Auschwitz conditions, etc." Thus an alternative or doubled self can live in a different world, an alternate reality, with its own set of rules and ethical conduct. The Auschwitz doctor does not deny reality, for he is "aware" that he is, for example, performing selections in the killing process, but he repudiates "the meaning of that reality." As an Auschwitz doctor, he does not believe that the selection process is murder. In addition, his original self repudiates and disavows "*anything* done by the Auschwitz self."[5] Through doubling then, the Nazi doctor is able to perform evil acts without believing or feeling that he is doing anything wrong.

The key to this situation, in my view as a philosopher and not a psychiatrist, is the establishment of a moral universe that is radically different from the moral universe of everyday reality. This is the context in which Lifton's doubled self will operate. The doubled self does not perceive his actions as evil, because they are in agreement with the standards of the new reality. Lifton claims that although each individual

Ethical Coda

Nazi doctor had his own style of doubling, "in all Nazi doctors, prior self and Auschwitz self were connected by the overall Nazi ethos and the general authority of the regime."[6] Nazi ideology created a new reality. This is then more than an issue in psychology; it is, rather, what we might call an issue in moral ontology. The moral agent comes to believe in a radically evil universe as good. One must act with self-justification in this radically evil context. Later I will return to the ideology of the Nazi worldview as it performs this ontological function of creating a new and evil reality for moral action—but now we must turn to the ideas and actions of technological professionals in the Third Reich.

Consider again Albert Speer, who began as Hitler's architect and rose to the highest level of the Nazi regime as Armaments Minister; eventually he was in charge of the entire industrial system of Germany. In Chapter Two, we touched briefly on his view of the neutrality of technology. It is worth reviewing here, and probing deeper into his justifications and rationalizations. As an architect, Speer is a technological professional, similar to an engineer, and thus he provides a useful case for the study of Nazi technological ethics. Speer was also the highest Nazi not killed during the war or executed after it; he was imprisoned, and wrote his memoirs.[7] This personal history is an essential resource for examining the moral ambiguities of the Nazi regime—even granting the fact that we may assume that everything Speer wrote after the war was a tool to promote his own interests.

Speer, as we saw, endorses a position of technological neutrality as the explanation for his evil acts. He claims that he was a pure technocrat unconcerned with ethical and political tasks. In commenting on his lack of concern for the virulent anti-Semitism of Hitler and the regime, he writes, "I felt myself to be Hitler's architect. Political events did not concern me … The grotesque extent to which I clung to this illusion is indicated by a memorandum of mine to Hitler as late as 1944: 'The task I have to fulfill is an unpolitical one. I have felt at ease in my work only so long as my person and my work were evaluated solely by the standard of practical accomplishments.'"[8] Let us put aside the fact that in 1944 the supposedly "unpolitical" practical tasks of Speer involved control of the entire wartime economy. The point is that Speer is attempting to create a distinct technological and practical realm that can be considered as separate from the realm of political and moral value. Technology is

Chapter Six

morally and politically neutral. As only an architect, involved with the design and creation of buildings, urban plans, and other artifacts, Speer cannot be concerned with the political and moral value of the things he produces for the master he serves.

If we adopt Lifton's perspective in our analysis of Speer's technological ethics, we can see a kind of doubling effect. Note in the quotation above that Speer only feels at ease in his politically neutral work, as if he had a second technological self: "so long as my *person* and my work were evaluated solely by the standard of practical accomplishments." But Speer's rationalization goes well beyond the existence of some kind of psychological doubling and into a critical evaluation of the nature of the technological professions. In explaining how his managerial style led to the renewed success of the armaments industry, Speer notes that he had placed technical people in control of his various programs: "I exploited the phenomenon of the technician's often blind devotion to his task. Because of what seems to be the moral neutrality of technology, these people were without any scruples about their activities."[9] It is not just Speer the architect who is unconcerned with political and moral values; according to Speer, technological professionals, embedded in a world of neutral technological artifacts, are blind to the normative dimensions of their work and their products.

In Chapter Two, I claimed that Speer's rationalizations are a prime example of Langdon Winner's account of the "traditional" view of the neutrality of technology, viz., that in any normative analysis of technology the design and creation of a technological artifact must be separated from its use.[10] I will return to that discussion in the conclusion of this chapter. Here I want to argue that the traditional view of the neutrality of technology is also an example of moral ontology, i.e., the creation of a particular moral universe. By making a clear and hard distinction between the design/creation and the use of a technological artifact, we are essentially establishing two normative realms. Obviously, there is only one physical artifact, but its meaning, and its subsequent value, will be different when examined in the two different spheres of reality. From the perspective of design, we will examine a gun, for example, in light of those various characteristics that make it an efficient gun—weight, balance, ease of use, etc. But from the perspective of use, we will examine the gun in light of the purposes and goals of the gun,

Ethical Coda

and the attainment (or not) of those goals—shooting a hunted animal, defense of a home, killing an enemy combatant, etc. There are two different sets of meaning and value for the gun, for the artifact. Thus, two normative realms are created when we adopt the traditional view of the neutrality of technology; this is, again, an expression of moral ontology.

How do the Nazi engineers fit into this picture of moral ontology? The designers of the crematoria ovens are a good example. As I reviewed the history in Chapter Two, the industrial furnace company of Topf and Sons was a major developer of the efficient multi-person crematorium ovens used at the SS controlled concentration camps. These ovens were originally planned for the prisoners who died from "natural" causes—malnutrition, disease, overwork, or punishment. As historians Jean-Claude Pressac and Robert Jan van Pelt recount this story, beginning in the late 1930s, the design and construction of the ovens for the camps were significantly different than crematoria furnaces built for commercial funeral establishments.[11] First, the ovens lacked any conventional aesthetic features, since there would be no family of the deceased to witness a ceremonial burning of the corpse. But more importantly, innovations in furnace technology permitted higher capacity and more efficient burns in the ovens, so that ovens could be designed to hold two or more bodies at the same time. There was no need to preserve the integrity of the individual ashes; there would be no bereaved family collecting the remains. Throughout the history of the design and construction of the crematoria ovens that were eventually to be used at Auschwitz and Birkenau, we find ever-increasing chambers for the incineration of corpses, from two to three to a double-furnace with four chambers each. The more efficient capacity for the ovens was a necessary requirement for the implementation of the Final Solution. The increased capacity of the ovens meant that the SS could handle the increased load of the direct killing operations.

From the perspective of the moral ontology of the traditional view of a neutral technology the engineers who designed and built these ovens could focus solely on the design problems with little or no regard for the ultimate uses of the artifacts. Engineers from Topf and Sons even came to Birkenau to deal with technical problems, such as the cracking of the smokestacks and uneven heat transference in Crema IV. Beyond the furnaces, there were serious problems with the ventilation and exhaust

Chapter Six

systems that were required for operational gas chambers. In all of these activities, professional engineers from Topf and Sons or the SS used their best technical expertise to create the necessary technological artifacts. These technical professionals seem to be following the dictum of Speer that the technician has a blind devotion to his practical task, without any concern for moral scruples.

Although Speer himself as far as we know had no connection to the Auschwitz killing center or any of the other death camps—Speer was sentenced at Nuremberg only for crimes related to the slave labor camp at the Dora-Mittelbau missile works—he provides a defense of his actions, and implicitly a defense of the actions of any technological professional involved in morally questionable projects. Speer's defense has two parts—both of which we examined in Chapter Two. First, he claims that in totalitarian political systems isolation and secrecy prevent a technological professional from being aware of the applications of the technology.[12] But in the second part of his defense Speer cunningly refuses to use this isolation as exculpation for his guilt. Instead he blames the traditional evaluation of technology as neutral. "It is true," he writes, "I was isolated. It is also true that the habit of thinking within the limits of my own field provided me, both as an architect and as Armaments minister, with many opportunities for evasion."[13] Thus he claims that he should have known about the evils of the Final Solution, and his moral failing is that he did not overcome his self-imposed isolation from political events. Speer clearly accepts the traditional view of the neutrality of technological artifacts by acknowledging that one could simply think and act "within the limits of [one's] own field," that is, one can ignore the political and moral realities of the technological project with which one is engaged. He is morally guilty because he did not escape the limitations of his technological thinking.

Legal scholar Jack Sammons has called Speer's position the endorsement of "rebellious ethics."[14] According to Sammons, one adopts the position of "rebellious ethics" when one claims that to be ethical one must rebel against the expectations and practices of one's profession. Indeed, for Sammons, this "is the dominant paradigm for the ethics of our professions." It means that "as ethical people ... we must stand apart from our professional roles in personal moral judgment of them."[15] Clearly, this is the narrative that Speer tells us in his memoirs.

Ethical Coda

Sammons calls it the position of the "Pure Technician ... the expert who is not accountable beyond his area of expertise. His technique, he claims, is morally neutral and he asks to be judged only by whether his means are the most efficient ones toward whatever end is given to him."[16] But the role of the pure technician leads to moral corruption, as we become consumed, as Speer and the Nazi engineers, with the task at hand regardless of the human cost. The only way to avoid this moral corruption, according to this paradigm, is to separate ourselves from our profession: "we must consciously maintain a personal and psychological detachment from our professional roles."[17]

The position of rebellious ethics then is clearly connected to the moral ontology that I have been developing here: there are two separate moral realms, and it is the task of the ethical technological professional to avoid being captured by the realm of moral neutrality. He can accomplish this by a kind of doubling, or re-asserting his personal moral self in rebellion against the demands of his profession. But Sammons rejects the paradigm of rebellious ethics, and with it, Speer's defense of his moral failure. For Sammons, Speer did not fail as a moral person because he failed to rebel against his professional role; rather he failed as a moral person because he failed *in his professional role* as an architect. It is a deeper integration with one's professional role that can provide a person with the moral resources to resist the evil practices of technology.[18]

What does this further integration mean? Sammons argues that the various technological professions provide us with moral guidelines that are built into the very nature of the profession itself. Architecture, for example, should be based on the idea that we are constructing built environments for human beings to live fuller and more creative lives, not buildings or cities that oppress and dominate the human spirit. Sammons notes that Speer's only moments of ethical insight were when he saw the morally evil directions of Hitler's ideas about architecture:

> As an architect, Speer began to see in Hitler's obsession with huge dimensions, his "violation of the human scale," his lack of proportion, his lack of concern for the social dimensions of architecture, his use of architecture as only an expression of his strength, and the pomposity and decadence of the style, a dictator bent on world domination for the sole purpose of his own glorification.[19]

Thus, for Sammons, architecture itself "offered Speer a truer perspective on Hitler." It was in thinking about architecture that Speer could have

Chapter Six

developed the moral vision to see what Nazism really was.[20] Speer then is not a Pure Technician who failed as a moral agent because he did not rebel against the neutrality of technology; rather he is a Failed Architect, and his moral failure is that he rejected the positive human goals of his craft in the pursuit of money, fame, power, and self-glorification. The good architect, as well as any good technological professional, must find within the profession the moral principles that can provide the foundation to pursue one's technological goals in a morally positive way.

III.

Regarding moral ontology, Sammons's rejection of the model of rebellious ethics, with its argument for the further integration of ethical and professional values, suggests that there are not distinct moral realms. Technological practice and ethics exist together in a unified worldview. But what if this unified moral ontology is itself evil? The work of historian Michael Allen, which we also briefly surveyed in Chapter Two, provides us with a disturbing answer to this question.[21] Through his analysis of SS industrial policy, he argues that among SS managers and engineers there was a convergence between professional goals and political values. Nazi engineers believed that what they were doing was good: there was no need to rebel against the pure technique of their profession, nor was there a need to find another ground of value to resist the evils of Nazism.

Allen's claims are based on his study of the lives of thirty-nine members of the elite SS engineering corps under the directorship of Hans Kammler, the Chief Engineer of the SS—a number comprising two-thirds of the elite corps.[22] Here is the story of Kurt Wisselinck,[23] the Chief Factory Representative within the SS to the German Workers Front or DAF, which had replaced all German labor unions once the Nazi Party seized power. His mid-level managerial position was situated at the confluence of several conflicting political power centers within the Third Reich: the SS, the DAF, the WVHA (the SS Building Division), and thus we might expect that the decisions Wisselinck made would tend to favor a particular bureaucratic allegiance. But what Allen discovers instead is that Wisselinck acted according to strict ideological principles, even when this ideology worked against the interests of major constituencies. The primary example is Wisselinck's handling of misconduct at the SS Granite Works of Gross-Rosen. What was the

Ethical Coda

misconduct? Not, one might suppose, that the prisoner-laborers were being worked to death; rather the problem was that the clothing of the Jewish prisoners was not being distributed to SS manager-trainees, and that one of the cooks was favoring the prisoners by giving them extra potatoes in their rations. The extra potatoes, of course, could mean the difference between life and death for the slave laborers, but more importantly, from the perspective of the SS business operation, the extra rations would make more efficient and productive workers. The managers at Gross-Rosen even complained that Wisselinck's presence was impeding the efficiency of the factory operations.

Wisselinck did not care about the success of the business operations. For him, as we learn in a memo he wrote to himself after his return from Gross-Rosen, the main point was the furtherance of Nazi ideology. He wrote: "The business undertakings of the Schutzstaffel [SS] are the best means to breathe new life into National Socialist ideals, to let them become reality ... Our example must spur other corporations forward to emulate us in order to see the growth of a happy, satisfied, and happy *Volk*."[24] Wisselinck is emblematic of the professional manager and technocrat within the Nazi regime, where consensus was built because of a shared ideology. The Nazi business and technical operations were efficient because "ideals, individuals, and institutions reinforced each other."[25]

As we noted in Chapter Two, Allen considers five main ideas as the basis of the shared ideology of SS managers and engineers.[26] Let us review them here: (1) a belief that the SS were the vanguard of a New Order that would "remake Europe in its own image;" (2) a commitment to the Führer or leadership principle; (3) a commitment to producing authentic German culture and values through the operation of business and technology, rather than a commitment to profit or wealth (the latter goals were obviously "Jewish"); (4) a fascination with modern technological organization, as represented by Fordism and Taylorism; and (5) biological racism and anti-Semitism . Whether these five principles are the sum total of the SS Nazi ideology is, of course, not the issue for me. Historians can debate the precise number of Nazi ideals and their relative importance. For my argument, it is necessary only to acknowledge that something like this set of principles existed as the basis of a Nazi worldview. Taken together, these principles form a coherent ideology, and indeed, as I indicated above, the basis of a distinct moral ontology.

Chapter Six

What this means is that the members of the Nazi SS engineering and management corps shared a set of overarching values that informed all of their decisions and actions. As Allen writes: "Ideology facilitated operations precisely because the maintenance of consensus never needed to be a heated topic of daily declarations and contention. It had become a matter of their collective identity as engineers of the New Order."[27] In short, these technological professionals believed in what they were doing. In Allen's wonderful description, they "were the model citizens of a murderous regime."[28] The Nazi ideology was their moral stance regarding the world. Thus there was no need for the psychological process of doubling; their personal worldview and the Nazi worldview were identical. This conclusion has important implications for the ethical evaluation of technology.

IV.

We have reached a very dark place. Let me review the argument before I arrive at a conclusion. When we consider the actions of technological professionals in the Nazi regime, we are drawn to the traditional view that the creation of technological artifacts is ethically neutral, and that only the use of these artifacts should bear a normative analysis. This view of technology can provide a Nazi engineer with a rational basis for the disavowal of guilt, and indeed, for a declaration of moral purity. It is almost identical to the psychological process of doubling which Lifton claims to have discovered in his analysis of the Nazi doctors, or to the model of rebellion against the ethics of the Pure Technician that Speer (belatedly) espouses in his memoirs.

Sammons has suggested that the correct alternative to rebellious ethics or Lifton's doubling is the further integration of personal and professional morality, rather than its separation. The technological professional should embrace the ethical foundations of his profession, for in those principles the professional can find the resources to resist the unethical or destructive forces that may corrupt or subvert his ideals. Architecture creates built environments for the betterment of human life; engineering creates technological artifacts for increased efficiency, comfort, and convenience. But what if the engineer or technician does not see the cultural and political forces that guide the profession in a particular historical context as a corruption or subversion? What if the

Ethical Coda

architect sees coherence between the political values of a murderous regime and the ideals of his profession? This is the lesson we learn from Allen's analysis of the SS engineers and industrial managers. A synthesis of Allen's analysis and Sammons' argument leads us to the following conclusion: for these technological professionals there was no need to rebel against one's professional technological ideals; there was no need to claim that technology is ethically neutral. In the Nazi regime the technology served both the purposes of the state and the ethical values of the technological professionals.

The technology of the Nazi state serves as a prime illustration of Winner's thesis that was discussed in Chapter Two that artifacts embody political and social values.[29] Winner criticizes the traditional view of the neutrality of technology by casting doubt on the strict separation of the "making" (which includes design and creation) and the "use" of technological artifacts. If this separation has any normative weight, it would permit the traditionally minded engineer to claim that his task—design and creation, i.e., making—is ethically neutral, and that he is thus blameless in any potential evil uses of the artifact. But Winner argues cogently that the distinction of making from use has no normative significance. The distinction rests on an overemphasis of the idea that technology is a mere tool of human activity. Winner instead argues that technological artifacts are "forms of life." Technologies become embedded in human life. Technological artifacts profoundly restructure and reshape human life. Winner writes: "As they become woven into the texture of everyday existence, the devices, techniques, and systems we adopt shed their tool-like qualities to become part of our very humanity."[30] This more robust idea of the role of technology in human life requires a more comprehensive method of evaluating the value of artifacts and technological systems. Once we realize that technologies alter human life we will see that they are not neutral tools but value-laden systems that create new forms of human reality. Technological artifacts actualize particular moral ontologies.

Winner has a short hand phrase for his view of technology: for him, "artifacts have politics." But I view Winner's conception more broadly to include ethics and social and cultural norms—for me, all of these values are embedded in technological artifacts and systems from the moment of their creation. We can see this from a practical point of

Chapter Six

view by examining the idea of "purpose." Any technological artifact must have a purpose before it is designed and created. Without such a purpose, there would be no reason to create the technology. But purpose implies a value; there are no neutral purposes. Thus the traditional separation between the making and the use of a technology cannot be the basis of a distinction in the existence of normative value. Normative value is omnipresent, in the initial conception and design of a technology, as well as its creation and use.

Here is the crux of the analysis of the Nazi engineers. For the technological professionals who believed in the Nazi ideological worldview, the design of technological artifacts and systems was a means to advance the values of National Socialism. Purpose and design went together. There was no possibility of neutrality, no possibility of what Speer called "the technician's blind devotion to his task." These technicians, these engineers, obviously believed in their task. Return to the historical narrative of the crematoria ovens at Auschwitz-Birkenau. The increased capacity of the internal oven chambers and the design that obviated the need for maintaining the integrity of the remains was a clear indication that the ovens would be part of a technological system of mass murder. Could this be a neutral design that was simply mis-used by evil concentration camp administrators? Such an idea is laughable. Pressac and van Pelt clearly address this point in their history of the design of the gas chambers and ovens that we reviewed in Chapter Two. It was during the planning phase of the new building at Birkenau, which would supplement the existing unit already operational at the Auschwitz main camp that "slowly the men in the WVHA [the SS Building unit] [began] to associate the 'final solution of the Jewish problem' with the capacity of the new crematorium."[31] Moreover, at an important engineering meeting on August 19, 1942, at Auschwitz, the chief engineer from Topf and Sons and a leading smokestack expert met with the SS engineers. They discussed the special actions and special works that would be necessary for these new buildings at Birkenau. As Pressac and van Pelt conclude, "it was clear to all participants in this meeting that crematoria IV and V would be involved in mass murder."[32] In short, the engineers did not design a neutral artifact; rather, they knowingly created a technological system for the purpose of genocide. The technological system would solve

Ethical Coda

a crucial political and social problem of the Nazi regime, and further what the SS engineers believed to be a necessary improvement in human life.

What does this mean for us and for my students who will be future engineers and technological professionals? If all technological artifacts have values embedded within them, then whatever engineers create will embody a particular set of political, ethical, social, and cultural norms and ideologies. I know that my engineering students fervently believe that the things that they will create will lead to a better world. But they also believe—somewhat paradoxically—that the artifacts that they create are value-neutral until they are used, that it is the end-user that bears the ultimate responsibility for the ethical value of the technology. How does this study of the Nazi engineers help the ethical development of engineers and technological professionals?

There are two issues here: (1) the idea of the value-neutrality of technological artifacts; and (2) the embodiment of positive social values in technological artifacts. Regarding the first issue, clearly the study of the history and ideology of the Nazi engineers reveals that the idea of a value-free technology is an illusion. By reviewing the narrative of the creation of the Nazi death camps we learn that technological artifacts are not value neutral, since they can be created solely for the purpose of fostering genocide. Thus the second issue comes to the foreground: how can engineers know that the technologies that they create embody positive social values that will improve the world? Perhaps they—and we—are in the grip of a misguided ideological worldview that embodies values fundamentally detrimental to human life. Many environmentalists and critics of consumerism, for example, might make this claim. Similarly, many critics of weapons development or ancillary technologies such as robotics and artificial intelligence might also. Our technologies may be leading to the death and destruction of more human life and even of the earth and its natural processes. How can we know?

We thus arrive at what I called above the most fundamental question for an understanding of engineering ethics: how does an engineer know that the values he embodies through his technological products are good values that will lead to a better world? I cannot answer this question in a satisfactory way. A proper answer would bring us to the fundamental question of all ethical thought: how do I know that my actions lead to the good? But philosophy's failure to answer these fun-

Chapter Six

damental ethical questions does not mean that the study of the Nazi engineers is useless. It is true that I cannot present my students with an answer to the fundamental question of engineering ethics. But I can show them, following the argument of Sammons, that their chosen profession embodies an ideal of social value that can be used as a moral guide. The danger, as revealed by the historical analysis of Allen, is that the ideal can be perverted by corrupt political and social ideologies. To live and work as ethical engineers, my students must be aware of the political and social goals that are served by their technological products. I can thus claim that if we remain in the thrall of the traditional view that the design and creation of technology is ethically neutral, then we will be repeating the mistakes of the Nazi engineers. We will be like that exemplar of Nazi technological efficiency, Albert Speer, who chose to ignore the entire political and social context of his professional tasks. To avoid the technological ethics of Hell we must always consider the normative purposes of our technological projects.

Can this analysis of the ethical obligations and misdeeds of engineers help us to understand the ideal of resistance to domination symbolized by Anne Frank's tree? In what ways can we interpret the ethical lessons here as a means to fight for liberation and autonomy, and against oppression and domination? Perhaps we need look no further than Sammons' concept of a professional ideal as the guiding ethical principle for those in technological fields. Perhaps Anne's tree can symbolize, not only a free and autonomous natural world, but also an ideal of professional behavior, free of the political and social goals of particular regimes and cultures. The tree reminds us to do good, to fight oppression, to strive for freedom, no matter what oppressive forces surround us. The tree, as a part of the natural world, will continue to exist and thrive, despite all the horrors and evils perpetuated by human history. The tree reminds us, in Anne words, "that everything will change for the better, that this cruelty too will end, that peace and tranquility will return once more."[33] It is to this ideal that we all must turn to find the courage to resist the forces of domination.

Ethical Coda

Notes to Chapter Six

1. Lifton 1986.
2. Lifton 1986, 418.
3. Lifton 1986, 420.
4. Lifton 1986, 421, original emphasis.
5. Lifton 1986, 422, original emphasis.
6. Lifton 1986, 425.
7. Speer 1970.
8. Speer 1970, 112.
9. Speer 1970, 212.
10. Winner 1986; Katz 2005.
11. Pressac and van Pelt 1994.
12. Speer 1970, 112–13.
13. Speer 1970, 113.
14. Sammons 1992.
15. Sammons 1992, 77.
16. Sammons 1992, 79.
17. Sammons 1992, 80.
18. Sammons 1992, 81.
19. Sammons 1992, 82–83.
20. Sammons 1992, 83.
21. Allen 2002.
22. Allen 2002, 159.
23. Allen 2002, 7–11.
24. Allen 2002, 9–10.
25. Allen 2002, 11.
26. Allen 2002, 12–16.
27. Allen 2002, 164.
28. Allen 2002, 5.
29. Winner 1986.
30. Winner 1986, 12.
31. Pressac and van Pelt 1994, 216.
32. Pressac and van Pelt 1994, 219.
33. Frank 1991, 333: 15 July 1944.

~ Epilogue ~

FIRE ISLAND, JULY 2012

The Atlantic Ocean is rough, although the day is fair and sunny, and there is a comfortable breeze blowing off the water. The waves are crashing far out, and a good size wash, about a foot deep is rolling up the beach. I am sitting in the middle of this wash, my legs splayed out, and holding in my lap a three-year old boy named Milos, the son of my friend Andrew who is visiting Fire Island. Earlier, Milos had appeared frightened of the water, and had refused to go in, but he gladly accompanied me into the surf to sit on my lap and have the wash from the waves roll over us, up to our waists. He still seems a bit concerned, especially each time a wave crashes, but I am holding him tight and he is having fun when each new batch of water hits us. His little sister Marika, all of eighteen months, is running around on the sand behind us, laughing at the waves or the birds or at nothing at all, under the watchful gaze of their father. Andrew is amazed that Milos went into the ocean with me; it is the boy's first time in the ocean waves.

And I am amazed at the mere presence of Milos and Marika, for given the currents of history that have been touched upon in this book, they should not even be alive, much less here, with me, at the ocean shoreline. Their maternal grandmother is a Holocaust survivor, who was hidden at the age of six with a Christian family in Budapest. She managed to avoid the mass deportations of the Hungarian Jewry in the spring of 1944, a series of deportations that ended at Auschwitz-Birkenau. If she had not been hidden, if she had been deported on the freight cars that traversed the Eastern European landscape, she would have died within an hour of her arrival at the camp at Birkenau, gassed, and her body incinerated to ashes in the crematorium ovens. Yet she did not die, indeed she is still alive, and she can watch her grandchildren play in the ocean surf.

Sometimes it seems to me that the existence of any Jewish children, after the Holocaust, is a miracle, a reality that defies explanation. But of course not all Jews were threatened by the Nazi regime, as I realized in the Spanish synagogue of Venice. American Jews after all never faced

Fire Island, July 2012

the danger of the death camps. So there is something special about these two children who escaped the evil causality of European history. They are an embodiment of that better world, filled with peace and tranquility, that captured the hopes of Anne Frank.

And they are here at the shores of the Atlantic Ocean, the physical place that means more to me than any other location on earth. For me, and for the themes of this book, it is significant that we are at the ocean, a natural entity that in its autonomy mocks the idea that we can control nature. Although the focus of this book has been a particular tree that existed in a precise urban landscape, and the forests and natural world that it represents, in my life it is always the ocean that signifies the wonder of a natural world beyond the boundaries of humanity. We humans can attempt to control the amount of erosion that the ocean causes on the beach, and we do, by erecting sand fences, planting beach grass, and even scraping the excess sand of the lower beach to build higher dunes—but there is no way we control the actual ocean. We will all be reminded of this truth, once again, in a few months when the winds and storm surges of Hurricane Sandy lay waste to the very beach on which we are sitting this summer day. To understand the meaning of a liberated and autonomous nature, all one needs to do is to gaze at the sea.

So here on this one day in July I confront in real life the ideas that I have presented in this book: the Holocaust survivor who defied the system of oppression and the autonomous power of a free and independent nature. Sitting with Milos I am consumed by the memories of an historical past, but the continuous presence of the ocean—of the natural world—reminds me that the life of humanity and nature is, in some sense, eternal. And memory seems to be the key to my understanding of this relationship, as it was in the Warsaw cemetery, the Majdanek death camp, the Spanish synagogue, the streets of Jewish Barcelona, or the Anne Frank house. Robert Pogue Harrison, whose analysis of the human connection to the forest we examined earlier writes, "to be human means to dwell in the openness of time, in defiance of the oblivion of nature, and hence to be governed by memory."[1] Nature creates an oblivion for humans, because the physical self must die, and with it, all personal consciousness of the world. But this oblivion is conquered through memory—not our memory, for we are dead, but the memories of those who come after us. Harrison concludes that memory "does not

Epilogue

merely look backward ... it also preempts the future." To be human means to look "forward to the memory of future generations."[2]

What will future generations think of us and our attempts to dominate both humanity and nature? I hope that they will find that there were enough voices raised in opposition to resist the forces of oppression, voices like those of Anne Frank contemplating the meaning of her tree. The Anne Frank tree has been a symbol for me of a free and independent nature and of the capacity for human individuals to resist the evils of manipulation, control, and domination. But here, in this place, I do not need any symbols. I know that I can sit here on the beach and think of Milos and Marika as I watch the ocean waves. The ocean is not a symbol, but actual physical nature, forever uncontrolled by humanity; and these are real-life children playing in the surf and sand. Their existence is a final rebuke to the evils of history and the misguided attempt by humanity to dominate the entire natural and human world.

Milos gets up from my lap and runs to play with his sister. They sit at the edge of the water and begin to build a sand castle. If I were in a schule, a synagogue, I would cover them with my prayer shawl, so that they could receive the blessing of the generations.

Notes to Epilogue

1. Harrison 1992, 13
2. Harrison 1992, 14.

References

Allen, Michael Thad. 2002. *The Business of Genocide: The SS, Slave Labor, and the Concentration Camps*. Chapel Hill: University of North Carolina Press.

"Arts Beat". 2011. "Stephen Sondheim Takes Issue with Plan for Revamped 'Porgy and Bess,'" *The New York Times*, August 10, 2011 [http://artsbeat.blogs.nytimes.com/2011/08/10/stephen-sondheim-takes-issue-with-plan-for-revamped-porgy-and-bess/]. Accessed March 25, 2015.

Bassin, Mark. 2005. "Blood or Soil? The *Völkisch* Movement, the Nazis, and the Legacy of Geopolitik," in Franz-Josef Brüggemeier, Mark Cioc, and Thomas Zeller (eds) *How Green Were the Nazis? Nature, Environment, and Nation in the Third Reich*. Athens, OH: Ohio University Press, pp. 204–242.

Berthold-Bond, Daniel. 2000. "The Ethics of 'Place': Reflections on Bioregionalism," *Environmental Ethics* 22: 5–24.

Birch, Thomas. 1990. "The Incarceration of Wildness: Wilderness Areas as Prisons," *Environmental Ethics* 12: 3–26.

Bramwell, Anna. 1985. *Blood and Soil: Richard Walther Darré and Hitler's 'Green Party'*. Bourne End: The Kensal Press.

Bramwell, Anna. 1989. *Ecology in the 20th Century: A History*. New Haven: Yale University Press.

Brennan, Andrew. 1988. *Thinking About Nature: An Investigation of Nature, Value, and Ecology*. Athens: University of Georgia Press.

Brennan, Andrew. 2006. "Globalization, Environmental Policy and the Ethics of Place," *Ethics, Place and Environment* 9: 133–148.

Brook, Isis. 2003. "Making Here Like There: Place Attachment, Displacement and the Urge to Garden," *Ethics, Place and Environment* 6: 227–234.

Brüggemeier, Franz-Josef, Mark Cioc, and Thomas Zeller (eds). 2005. *How Green Were the Nazis? Nature, Environment, and Nation in the Third Reich*. Athens, OH: Ohio University Press.

Closmann, Charles. 2005. "Legalizing a *Volksgemeinschaft*: Nazi Germany's Reich Nature Protection Law of 1935," in Franz-Josef Brüggemeier, Mark Cioc, and Thomas Zeller (eds) *How Green Were the Nazis? Nature, Environment, and Nation in the Third Reich*. Athens, OH: Ohio University Press, pp. 18–42.

Dawidowicz, Lucy S. 1975. *The War Against the Jews, 1933–1945*. New York: Holt, Rinehart and Winston.

Devall, Bill. 1988. *Simple in Means, Rich in Ends: Practicing Deep Ecology*. Salt Lake City: Gibbs Smith.

Duncan, Colin. A. M. 1991. "On Identifying a Sound Environmental Ethic in History: Prolegomena to Any Future Environmental History," *Environmental History Review* 15, no. 2: 5–30.

Eliot, T. S. 1925. "The Hollow Men," in T. S. Eliot, *Collected Poems*. New York: Harcourt, Brace Jovanovich, 1961.

References

Elliot, Robert. 1982. "Faking Nature," *Inquiry: An Interdisciplinary Journal of Philosophy* **25**: 81–93.

Elliot, Robert. 1997. *Faking Nature.* London and New York: Routledge.

Ellul, Jacques. 1964. *The Technological Society.* New York: Vintage Books.

Ellul, Jacques. 1983. "The Technological Order," in Carl Mitcham and Robert Mackey, (eds), *Philosophy and Technology: Readings in the Philosophical Problems of Technology.* New York: Free Press, pp. 86–105.

Evernden, Neil. 1992. *The Social Creation of Nature.* Baltimore: The Johns Hopkins University Press.

Florman, Samuel. 1975. *Existential Pleasures of Engineering.* New York: St. Martin's Press.

Frank, Anne. 1991. *The Diary of a Young Girl.* Edited by Otto H. Frank and Mirjam Pressler and translated by Susan Massotty. New York: Anchor Books.

Gilbert, Martin. 1985. *The Holocaust: A History of the Jews of Europe during the Second World War.* New York: Henry Holt.

Groening, Gert and Wolschke-Bulmahn, Joachim. 1992. "Some Notes on the Mania for Native Plants in Germany," *Landscape Journal* **11** (2): 116–126.

Haraway, Donna J. 1991. "A Cyborg Manifesto: Science, Technology and Socialist-Feminism in the Late Twentieth Century," in *Simians, Cyborgs, and Woman: The Reinvention of Nature.* New York: Routledge, pp. 149–181.

Harrison, Robert Pogue. 1992. *Forests: The Shadow of Civilization.* Chicago: University of Chicago Press.

Healy, Patrick. 2011. "'Porgy': No New Scene, Some Hard Feelings," *The New York Times*, November 15, 2011 [http://www.nytimes.com/2011/11/15/theater/the-gershwins-porgy-and-bess-is-less-changed-for-broadway.html]. Accessed March 25, 2015.

Hettinger, Ned. 2001. "Exotic Species, Naturalisation, and Biological Nativism," *Environmental Values* **10** (2): 193–224.

Hettinger, Ned. 2012. "Nature Restoration as a Paradigm for the Human Relationship with Nature," in Allen Thompson and Jeremy Bendik-Keymer (eds) *Ethical Adaptation to Climate Change: Human Virtues in the Future.* Cambridge: MIT Press, pp. 27–46.

Higgs, Eric. 2003. *Nature by Design: People, Natural Process, and Ecological Restoration.* Cambridge: MIT Press.

Hilberg, Raul. 2003. *The Destruction of the European Jews.* 3rd edition. New Haven: Yale University Press

Imort, Michael. 2005. "'Eternal Forest—Eternal *Volk*': The Rhetoric and Reality of National Socialist Forest Policy," in Franz-Josef Brüggemeier, Mark Cioc, and Thomas Zeller (eds) *How Green Were the Nazis? Nature, Environment, and Nation in the Third Reich.* Athens, OH: Ohio University Press, pp. 43–72.

Jaskot, Paul B. 2000. *The Architecture of Oppression: The SS, Forced Labor and the Nazi Monumental Building Economy.* London and New York: Routledge.

Jenkins, Willis. 2005. "Assessing Metaphors of Agency: Intervention, Perfection, and Care as Models of Environmental Practice," *Environmental Ethics* **27**: 135–154.

References

Jordan, William R., III. 1994. "The Nazi Connection," *Restoration & Management Notes* **12**: 113–114.

Katz, Eric. 1979. "Utilitarianism and Preservation," *Environmental Ethics* **1**: 357–364.

Katz, Eric. 1991. "Restoration and Redesign: The Ethical Significance of Human Intervention in Nature," *Restoration and Management Notes* **9**: 90–96.

Katz, Eric. 1992a. "The Big Lie: Human Restoration of Nature," *Research in Philosophy and Technology* **12**: 231–241.

Katz, Eric. 1992b. "The Call of the Wild: The Struggle Against Domination and the Technological Fix of Nature," *Environmental Ethics* **14**: 265–273.

Katz, Eric. 1993. "Artefacts and Functions: A Note on the Value of Nature," *Environmental Values* **2**: 223–232.

Katz, Eric. 1995. "Imperialism and Environmentalism," *Social Theory and Practice* **21**: 271–285.

Katz, Eric. 1996a. "Nature's Presence: Reflections on Healing and Domination," in Andrew Light and Jonathan M. Smith (eds) *Philosophy and Geography I: Space, Place, and Environmental Ethics*. Lanham, MD: Rowman & Littlefield, pp. 49–61.

Katz, Eric. 1996b. "The Problem of Ecological Restoration," *Environmental Ethics* **18**: 222–224.

Katz, Eric. 1997. *Nature as Subject: Human Obligation and Natural Community*. Lanham, MD: Rowman and Littlefield.

Katz, Eric. 1999. "A Pragmatic Reconsideration of Anthropocentrism," *Environmental Ethics* **21**: 377–390.

Katz, Eric. 2000. "Another Look at Restoration: Technology and Artificial Nature," in Paul H. Gobster and R. Bruce Hull (eds) *Restoring Nature: Perspectives from the Social Sciences and Humanities*. Washington D.C.: Island Press, pp. 37–48.

Katz, Eric. 2002. "The Liberation of Humanity and Nature," *Environmental Values* **11**: 397–405.

Katz, Eric. 2005. "On the Neutrality of Technology: The Holocaust Death Camps as a Counterexample," *Journal of Genocide Research* **7**: 409–421.

Katz, Eric. 2009. "Convergence and Ecological Restoration: A Counterexample," in Ben Minteer (ed.) *Nature in Common? Environmental Ethics and the Contested Foundations of Environmental Policy*. Philadelphia: Temple University Press, pp. 185–195.

Katz, Eric. 2010. "Anne Frank's Tree: Thoughts on Domination and the Paradox of Progress," *Ethics, Place and Environment* **13**: 283–293.

Katz, Eric. 2011. "Preserving the Distinction between Nature and Artifact," in Gregory E. Kaebnick (ed.) *The Ideal Of Nature: Debates about Biotechnology and the Environment*. Baltimore: Johns Hopkins University Press, pp. 71–83.

Katz, Eric. 2012. "Further Adventures in the Case Against Restoration," *Environmental Ethics* **34**: 67–97.

Keeling, Paul M. 2008. "Does the Idea of Wilderness Need a Defence?" *Environmental Values* **17**: 505–519.

References

Kidner, David. 2000. "Fabricating Nature: A Critique of the Social Construction of Nature," *Environmental Ethics* **22**: 339–357.

King, Roger J. H. 2000. "Environmental Ethics and the Built Environment," *Environmental Ethics* **22**: 115–131.

Kranzberg, Melvin. 1991. "Technology and Human Values," in William B. Thompson, (ed.), *Controlling Technology: Contemporary Issues*. Buffalo: Prometheus, pp. 157–165.

Krieger, Martin H. 1973. "What's Wrong with Plastic Trees?" *Science* **179**: 446–455.

Leiss, William. 1974. *The Domination of Nature*. Boston: Beacon Press.

Lekan, Thomas. 2005. "'It Shall Be the Whole Landscape!' The Reich Nature Protection Law and Regional Planning in the Third Reich," in Franz-Josef Brüggemeier, Mark Cioc, and Thomas Zeller (eds) *How Green Were the Nazis? Nature, Environment, and Nation in the Third Reich*. Athens, OH: Ohio University Press, pp. 73–100.

Levi, Primo. 1987. *The Reawakening*, translated by Stuart Woolf. New York: Collier Books.

Lewis, C. S. [1947] 1983. "The Abolition of Man," in Carl Mitcham and Robert Mackey, (eds) *Philosophy and Technology: Readings in the Philosophical Problems of Technology*. New York: Free Press, pp. 143–150.

Lifton, Robert J. 1986. *The Nazi Doctors: Medicalized Killing and the Psychology of Genocide*. New York: Basic Books.

Light, Andrew. 1998. "On the Irreplaceability of Place," *Worldviews: Environment, Culture, Religion* **2**: 179–184.

Light, Andrew. 2000. "Ecological Restoration and the Culture of Nature; A Pragmatic Perspective," in Paul H. Gobster and R. Bruce Hull (eds) *Restoring Nature: Perspectives from the Social Sciences and Humanities*. Washington D.C.: Island Press, pp. 49–70.

Light, Andrew. 2002. "Place Authenticity as Ontology or Psychological State? A Reply to Katz," *Philosophy & Geography* **5**: 204–210.

Lo, Yeuk-Sze. 1999. "Natural and Artifactual: Restored Nature as Subject," *Environmental Ethics* **21**: 247–66.

Malpas, Jeff. 1998. "Finding Place: Spatiality, Locality, and Subjectivity," in Andrew Light and Jonathan M. Smith (eds) *Philosophy and Geography III: Philosophies of Place*. Lanham, MD: Rowman and Littlefield, pp. 21–43.

Marcuse, Herbert. 1972. *Counterrevolution and Revolt*. Boston: Beacon Press.

Maser, Chris. 1988. *The Redesigned Forest*. San Pedro, CA: R& E Miles.

McKibben, Bill. 1989. *The End of Nature*. New York: Random House.

Merchant, Carolyn. 1980. *The Death of Nature; Women, Ecology, and the Scientific Revolution*. San Francisco: Harper and Row.

Mesthene, Emmanuel. 1983. "Technology and Wisdom," and "How Technology Will Shape the Future," in Carl Mitcham and Robert Mackey (eds) *Philosophy and Technology: Readings in the Philosophical Problems of Technology*. New York: Free Press, pp. 109–129.

Michael, Mark A. 2001. "How to Interfere with Nature," *Environmental Ethics* **23**: 135–154.

Michael, Mark A. 2005. "Is It Natural to Drive Species to Extinction?" *Ethics and the Environment* **10**: 49–66.

References

Moriarty, Paul Veatch. 2007. "Nature Naturalized: A Darwinian Defense of the Nature/Culture Distinction," *Environmental Ethics* **29**: 227–246.

Mumford, Lewis. 1963. *Technics and Civilization*. New York: Harcourt, Brace and World.

The New York Times August 24, 2012. "Despite Good Intentions, a Fresco in Spain Is Ruined"[http://www.nytimes.com/2012/08/24/world/europe/botched-restoration-of-ecce-homo-fresco-shocks-spain.html?_r=0]. Accessed January 10, 2015.

O'Brien, William. 2006. "Exotic Invasions, Nativism, and Ecological Restoration: On the Persistence of a Contentious Debate," *Ethics, Place and Environment* **6**: 63–77.

Packard, Steve. 1988. "Just a Few Oddball Species: Restoration and the Rediscovery of the Tallgrass Savanna," *Restoration and Management Notes* **6**: 13–22.

Peretti, Jonah H. 1998. "Nativism and Nature," *Environmental Values* **7**: 183–192.

Peterson, Anna. 1999. "Environmental Ethics and the Social Construction of Nature," *Environmental Ethics* **21**: 339–357.

Piper, Franciszek. 1994. "Gas Chambers and Crematoria," in Yisrael Gutman and Michael Berenbaum (eds) *Anatomy of the Auschwitz Death Camp*. Bloomington: Indiana University Press, pp. 157–182.

Pitt, Joseph. 2000. *Thinking About Technology: Foundations of the Philosophy of Technology*. New York: Seven Bridges Press.

Plumwood, Val. 1998. "Wilderness Skepticism and Wilderness Dualism," in J. Baird Callicott and Michael P. Nelson (eds) *The Great New Wilderness Debate*. Athens: University of Georgia Press, pp. 652–690.

Pollan, Michael. 1994. "Against Nativism," *The New York Times*, May 15, 1994 [http://www.nytimes.com/1994/05/15/magazine/against-nativism.html]. Accessed March 25, 2015.

Pressac, Jean-Claude, with Robert Jan van Pelt. 1994. "The Machinery of Mass Murder at Auschwitz," in Yisrael Gutman and Michael Berenbaum (eds) *Anatomy of the Auschwitz Death Camp*. Bloomington: Indiana University Press, pp. 183–245.

Preston, Christopher. 1999. "Environment and Belief: The Importance of Place in the Construction of Knowledge," *Ethics and the Environment* **4**: 211–218.

Rehmann-Sutter, Christoph. 1998. "An Introduction to Places," *Worldviews: Environment, Culture, Religion* **2**: 171–177.

Rodman, John. 1977. "The Liberation of Nature?" *Inquiry* **20**: 83–131.

Rolston, Holmes, III. 1988. *Environmental Ethics: Duties to and Values in the Natural World*. Philadelphia: Temple University Press.

Rolston, Holmes, III. 1998. "Down to Earth: Persons in Place in Natural History," in Andrew Light and Jonathan M. Smith (eds) *Philosophy and Geography III: Philosophies of Place*. Lanham, MD: Rowman and Littlefield, pp. 285–296.

Sagoff, Mark. 2005. "Do Non-Native Species Threaten the Natural Environment?" *Journal of Agricultural and Environmental Ethics* **18**: 215–236.

Sale, Kirkpatrick. 1985. *Dwellers in the Land: The Bioregional Vision*. San Francisco: Sierra Club.

References

Sammons, Jack. L. Jr. 1992. "Rebellious Ethics and Albert Speer," *Professional Ethics* 1 (3–4): 77–116.

Schama, Simon. 1995. *Landscape and Memory.* New York: Knopf.

Siipi, Helena. 2003. "Artefacts and Living Artefacts," *Environmental Values* 12: 413–430.

Siipi, Helena. 2008. "Dimensions of Naturalness," *Ethics and the Environment* 13: 71–103.

Singer, Peter. 1975. *Animal Liberation: A New Ethics for Our Treatment of Animals.* New York: Avon Books.

Smith, Mick. 1999. "To Speak of Trees: Social Constructivism, Environmental Values, and the Future of Deep Ecology," *Environmental Ethics* 21: 359–376.

Soper, Kate. 1995. *What is Nature? Culture, Politics and the non-Human.* Oxford: Blackwell.

Speer, Albert. 1970. *Inside the Third Reich: Memoirs*, translated by Richard and Clara Winston. New York: Simon and Schuster.

Stephens, Piers H. G. 2000. "Nature, Purity, Ontology," *Environmental Values* 9: 267–294.

Stone, Christopher. 1974. *Should Trees Have Standing? Toward Legal Rights for Natural Objects.* Los Altos, CA: William Kaufmann.

Sylvan, Richard. 1994. "Mucking with Nature," in *Against the Mainstream: Critical Environmental Essays.* Canberra: Australian National University, pp. 48–78.

Taylor, Bron. 2000. "Deep Ecology and its Social Philosophy: A Critique," in Eric Katz, Andrew Light, and David Rothenberg (eds) *Beneath the Surface: Critical Essays in the Philosophy of Deep Ecology.* Cambridge, MA: MIT Press, pp. 269–299.

Thompson, Janna. 2000. "Environment as Heritage," *Environmental Ethics* 22: 241–258.

Tuan, Yi-Fu. 1974. *Topophilia.* Engelwood Cliffs. NJ: Prentice-Hall.

Uekoetter, Frank. 2006. *The Green and the Brown: A History of Conservation in Nazi Germany.* Cambridge: Cambridge University Press.

United States Holocaust Memorial Museum. 1996. *Historical Atlas of the Holocaust.* New York: Macmillan.

van Pelt, Robert Jan. 1994. "A Site in Search of a Mission," in Yisrael Gutman and Michael Berenbaum (eds) *Anatomy of the Auschwitz Death Camp.* Bloomington: Indiana University Press, pp. 93–156.

van Pelt, Robert Jan. 2002. *The Case for Auschwitz: Evidence from the Irving Trial.* Bloomington: Indiana University Press.

Vogel, Steven. 1996. *Against Nature: The Concept of Nature in Critical Theory.* Albany: State University of New York Press.

Vogel, Steven. 2002. "Environmental Philosophy after the End of Nature," *Environmental Ethics* 24: 23–39.

Vogel, Steven. 2003. "The Nature of Artifacts," *Environmental Ethics* 25: 149–168.

Warren, Karen. 1990. "The Power and the Promise of Ecological Feminism," *Environmental Ethics* 12: 125–146.

Winner, Langdon. 1986. *The Whale and the Reactor: A Search for Limits in an Age of High Technology.* Chicago: University of Chicago Press.

References

Wolschke-Bulmahn, Joachim. 2005. "Violence as the Basis of National Socialist Landscape Planning in the 'Annexed Eastern Areas,'" in Franz-Josef Brüggemeier, Mark Cioc, and Thomas Zeller (eds) *How Green Were the Nazis? Nature, Environment, and Nation in the Third Reich*. Athens, OH: Ohio University Press, pp. 243–256.

Woods, Mark and Moriarty, Paul Veatch. 2001. "Strangers in a Strange Land: The Problem of Exotic Species," *Environmental Values*, **10**: 163–191.

Yahil, Leni. 1990. *The Fate of European Jewry*, translated by Ina Friedman and Haya Galai. New York: Oxford University Press.

Zeller, Thomas. 2005. "Molding the Landscape of Nazi Environmentalism: Alvin Seifert and the Third Reich," in Franz-Josef Brüggemeier, Mark Cioc, and Thomas Zeller (eds) *How Green Were the Nazis? Nature, Environment, and Nation in the Third Reich*. Athens, OH: Ohio University Press, pp. 147–170.

Index

A

abuse
  of environment; nature 9, 80, 124
  of power 31
aesthetic, the 2, 19, 48, 58, 61, 76, 93, 99, 126, 139, 143, 149, 159, 169
aesthetics 141, 152
affect (n.) 44
agent; agency 7, 9, 23, 36, 51, 81, 101–03, 157, 166, 167, 172
agriculture *see also* farmland 3, 11, 18, 50, 61, 73, 81, 83, 113, 126, 150, 156, 159
alien species 12, 141–9, 152
Allen, Michael 61–4, 172–5, 178
Allies; Allied Forces 15, 35
Amsterdam 1, 40, 73, 113, 132, 135
analogy 45, 131, 132, 153
  of Nature and human subject 7, 23, 27–9, 34, 77
  of ecological restoration and art forgery 73, 75
  of ecological restoration and procreation 90–1, 95
Anne Frank House 1, 2, 132–3, 181
anthropocentrism 9, 19–24, 29, 31, 46, 65, 82, 84, 88, 90, 94, 100, 117, 119, 124, 143
architect; architecture 18, 51, 55–7, 60–1, 67, 130, 152, 155, 164–8, 170–5
Aristotle 21, 127
art 47, 73–6, 121
artifact *see also* creation; design 4, 9, 21, 25–6, 35, 43, 45, 47–9, 50–5, 56–7, 60–2, 64, 66, 68, 72, 76–80, 81–102, 104–08, 113, 115, 126, 127, 129, 161, 164–5, 168–70, 174–7
Aryan 3, 62, 156, 160
Atlantic Ocean 32, 180–1

Auschwitz-Birkenau 10, 15, 56, 58–60, 166, 167, 169, 170, 176, 180
authenticity 11, 40–3, 47–50, 65–8, 75–7, 80–1, 121–2, 134, 138–9, 151, 155–6, 159, 161, 173
authority; authoritarian 31, 54–5, 121, 167
authorities 35
autonomy *see also* independence 1, 2, 4, 7–9, 12, 25–6, 28–36, 40, 42, 49, 72, 77–9, 81–2, 85–91, 95, 112, 116–20, 122, 124, 128, 133–4, 160–1, 178, 181
axiology 123

B

Bacon, Francis 27
Barcelona 68, 181
Bassin, Mark 156–8
bath 35
beach 25, 26, 32–4, 44, 71–2, 78, 180
  barrier 32
  replenishment; restoration 25–6, 32, 72, 78, 106, 108, 115
beauty 5, 6, 14, 16, 24, 35, 38, 40, 48, 66, 93–4, 98, 113, 123, 159
biodiversity 12, 141, 143, 144, 147, 148
bioregion; bioregionalism 42, 43, 46, 66, 148, 150
biotechnology 96
Birch, Thomas 106
birds 5, 6, 7, 32, 114, 133, 180
Bischoff, Karl 58
"blood and soil" 11, 139–40, 150–51, 156–7
Bonner, J.T. 104
Brennan, Andrew vii, 88
Brook, Isis 146–7
Brundtland report 20
Buchenwald 60
business 51, 62–4, 173

Index

C

category 23, 24, 25, 29–33, 36, 90, 96, 99, 114–16, 125, 126, 152
cemetery 2, 8, 14–16, 22, 24, 26–7, 34, 36–7, 40, 42, 49, 124–5, 181
child 31, 40, 41, 44, 67, 68, 89–91, 180–2
civilization 8, 18, 19, 36, 50, 88, 117, 121, 122, 130–1, 133, 134
climate change 18, 20, 113, 116, 154
Closmann, Charles 149, 151
community *see also* population
  ecological 142, 145, 146, 148
  human 7, 20, 25, 36, 39, 41, 42, 62, 67, 71, 72, 81, 94, 106, 107, 108, 121, 148, 151, 153–4, 156–9
conscience 166
consciousness 7, 28, 181
conservation *see also* preservation 20, 73, 79, 96, 116, 122, 143–4, 150–2, 154, 159
constructivism, social 11, 99, 113, 116–20, 123–9
continuity 8, 13, 23, 47–9, 61, 65–6, 74–6, 81–2, 133, 178, 181
corruption, moral 133–4, 171, 174, 178
Corsica 133–4
creation *see also* artifact; design 2, 4, 9, 10, 11–13, 18, 21, 24–7, 36, 43, 47–55, 60–5, 72, 75–84, 87–94, 97–102, 104, 106–07, 123, 125–9, 132, 139, 146–7, 149, 152–5, 157, 160–1, 165, 167–70, 174–8, 181
crematorium 13, 15–16, 50, 57–61, 65, 169, 176, 180
cultural construct *see also* constructivism 30, 103
culture; cultural 31, 40, 43–8, 51, 52, 62, 64, 66, 67, 68, 76, 78, 79, 81–3, 95, 103–05, 114, 116–21, 125, 126, 129–30, 141, 143, 145–9, 157, 158, 160, 173–5, 177, 178

D

Dachau 58
dam 31, 74
Dante 131, 134
Darré, Richard 11, 150, 156–7
Darwin; Darwinian 77, 88, 104
Dauerwald 12, 153
Dawidowicz, Lucy 41
death camp *see also* individual camp names 3, 5, 8, 10, 12, 14–16, 22, 27, 34, 36, 40–2, 49, 50, 52, 55, 57–8, 61, 66, 67, 72, 164, 166, 170, 177, 181
deep ecology 42–3
degradation, environmental 2, 9, 20, 26, 35, 47, 66, 73, 80, 125, 142–5
Descartes, René 131–2
design *see also* artifact; creation 12, 13, 18, 25, 26, 33, 35, 47, 51–61, 64–5, 75–8, 81, 83, 87–91107, 139, 152, 160, 164, 165, 168–9, 175–8
desire 21, 23–6, 35, 82, 123, 126, 128, 139, 145, 147–9, 151, 152
destruction 2, 6, 8–10, 14, 15–17, 19–22, 23, 24, 26, 27, 34–6, 40, 43, 48–9, 66, 67, 71–2, 78–80, 91, 95, 98, 123–5, 130, 131, 138, 143, 154–5, 174, 177
development 7, 23, 24, 26, 28, 31, 33–6, 44, 75–8, 79, 82, 89, 92, 130, 150, 151, 155, 161, 177
  economic 17–20, 32–3, 46, 49, 73, 74, 98, 115–16, 122–3, 133–4, 151
  sustainable 2, 9, 19–20, 122
  technological 51, 57, 65, 86, 169, 177
disaster, natural 23, 78
discourse 99, 100, 115–16, 126–7, 135
discrimination 31, 99, 126, 141
doctor, Nazi 165–7, 174
domination *see also* mastery; power 1–13, 17–36, 40, 50, 52, 57, 65–8, 72–3, 78–82, 91, 95, 105, 108, 112–15, 120–9, 131–5, 138–9, 149, 155–61, 164, 171, 178, 182

Index

Dora-Mittelbau 63, 65, 170
"doubling" (psychological) 165–8, 171, 174
dualism (of culture and nature) 10, 82–8, 91, 99–105, 2116, 117, 126–9
Duncan, Colin 23

E

Eastern Europe 2, 14, 17, 24, 34, 35, 41, 61, 124, 146, 154–5, 157–9, 160–1
"ecocide" 17, 22
ecofeminism 27
economy; economic 2, 18–20, 24, 31, 32, 73, 76, 119, 122–3, 133, 141, 143, 152, 159, 167
ecosystem 12, 20–1, 25, 31, 34, 35, 42, 45–7, 66, 72–4, 79–80, 85, 90–1, 98–9, 107, 123, 140–50, 153, 158
education 19, 31, 45, 165
elimination
  of people *see also* genocide 18, 62, 65, 146, 149, 154
  of other species 12, 74, 85, 140–1, 144, 145–6, 147, 149, 153, 156, 160
  of human influence on land 33, 79, 155, 160
eliminationism 12, 138–40, 146, 156
Eliot, T.S. 92
Elliot, Robert 73–6
Ellul, Jacques 51, 64–5
emotion 23, 24, 40, 42, 44, 454, 147
Ems, R. 159
endangered species 32–4, 145
Engels, Friedrich 54
engineer; engineering 2, 3, 12, 13, 51, 57, 58, 61, 67, 159, 164–5, 167, 169–78
Enlightenment 1, 18, 19, 50, 121
epistemology 11, 28–30, 34, 118, 123–5, 134, 138
erosion 25, 32, 71, 107, 181
ethics *see also* morality; values 31, 32, 36, 43, 47, 96, 102, 116–17, 138, 141, 164–78

environmental 27, 29, 32, 42, 43, 46, 66, 99, 103, 117, 159
evaluation 7, 9, 12, 21, 29, 36, 43, 44, 46, 50–1, 53–4, 66, 75–6, 93–4, 102, 105–07, 117, 167–8, 170, 174, 175
Evernden, Neil 127–9
evil 1–2, 4, 6–9, 11–12, 14–17, 22, 24, 26–7, 31, 34–6, 40, 42, 49, 51–3, 55, 57, 65, 105, 106, 112, 114, 117, 120, 135, 138, 150, 152, 157–8, 160–1, 165–7, 170–2, 175, 176, 178, 181–2
exotic species 12, 74, 140–9, 152, 156, 158, 160
extinction 140, 143–5, 153

F

farmland *see also* agriculture 14, 17, 21, 107, 159
fire 25, 106
Fire Island 2, 32–4, 44, 71–2, 78, 94, 107, 115, 180–2
Florman, Samuel 51
Flossenbürg 60–1, 65
Fordism 62, 173
forest 1–2, 8, 10, 11, 24–5, 31, 81–2, 97–8, 121–5129–34, 153, 160, 181
forestry 2, 11–12, 19, 113, 132, 151, 153–4
Frank, Anne 1, 4, 5, 6, 10, 28, 31, 32, 34, 36, 49, 105, 112, 133, 138, 181, 182
  tree 1, 3, 4, 7, 8, 11, 12, 17, 27, 28, 31, 34, 36, 40, 42, 49, 68, 72, 95, 121, 132–4, 138, 160, 164, 178, 182
freedom *see also* liberation 2, 7, 9, 30, 31, 33, 46, 49, 68, 77, 91, 121, 122, 127, 128, 155, 178
furnace *see also* crematorium 57–61, 65, 169

Index

G

garden; gardening 3, 80–2, 131, 139–41, 146–7, 149, 152, 160
gas chamber 13, 15, 50, 57–60, 170, 176
gender 31, 52, 63
genocide *see also* elimination 3, 9, 12, 13, 16–17, 21, 22, 24, 26, 41, 43, 52, 57, 65, 138, 149, 161, 164, 176, 177
Germany; German 17–18, 39, 41, 49, 60–64, 121–2, 132, 139, 140, 146–7, 150–9, 161, 165, 167, 172–3
Germanizing 18, 147, 154–5, 158–9
ghetto 14, 39, 50, 57, 66–8
global warming *see* climate change
globalization 12, 147–9
God 5, 19, 36, 77
good 3, 4, 6, 8, 13, 19, 20, 22–5, 29, 51, 53, 57, 62, 80, 93, 100, 113, 117, 119, 124, 133, 141, 158–60, 165–7, 172, 177–8
Göring, Herman 153
Gottlieb, Roger vi
government 20, 32, 146, 149
grave 2, 8, 9, 14, 26, 36, 49, 124
Greek 21, 121
green (environmentalist) 2, 105, 150
Gross-Rosen 172–3
guilt 21, 41, 56, 57, 166, 170, 174
Gulf of Mexico 24, 31
gun 51, 168–9

H

Habermas, Jürgen 29
happiness 5, 6, 18, 19, 90, 133, 173
Haraway, Donna 86
Hargrove, Eugene vi
harm 4, 20, 23–4, 31, 79–80, 93, 95, 141, 143–5, 158, 160
harmony 17, 62, 112, 143
Harrison, Robert Pogue 129–34, 181
Hassardi, Samuel 68
hatred 6, 14, 21, 56

healing 2, 8, 9, 16, 22, 24–7, 35–6, 42, 49, 80–2, 85
Hebrew 38, 40, 48, 68
Heck, Lutz 158
Hegel, Georg Wilhelm Friedrich ; Hegelian 23
Heimat 150–4, 157
Hengeveld, Rob 143
Herder, Johann Gottfried 121
heritage 45
Hettinger, Ned vi, 80, 144, 147–8
hiding 4, 5, 23, 49
Higgs, Eric vi, 79–81, 108
Himmler, Heinrich 18, 61, 154–5
Hitler, Adolf 18, 56, 57, 167, 171
Holocaust 4, 9, 14, 16–217, 21–2, 27, 35, 40–3, 48–50, 52, 66, 68, 82, 124, 132, 140, 180, 181
home 25, 32, 44–5, 66, 68, 71, 72, 113, 115, 139, 147–9, 153–6, 169
homeland see also *Heimat* 3, 18, 145–55, 157–8, 161
humanism 21, 128
humility 6, 92–3, 161
hybridity 32, 34, 72, 78

I

ideology 4, 10, 12, 13, 19–21, 61–5, 112, 118, 140, 147, 150–61, 165, 167, 172–8
imperialism 2, 23–4, 31, 35, 128, 141, 155, 156
independence
  cultural 130
  of nature *see also* autonomy 9–12, 24, 26, 27–9, 34, 36, 49, 68, 72, 95, 96, 108, 112–16, 120, 123, 125–7, 129, 133, 135, 181–2
  of technology *see also* neutrality 54
indigenous
  people 116–17, 147
  other species *see also* native; nativism 141
industry; industrial 18, 20, 63, 97, 98, 113, 122, 125, 150, 164–5, 167–9, 172, 175

Index

institutions
  human; societal 1, 7, 10, 11, 19, 23, 24, 29, 30, 46, 48, 53, 54, 106, 124
  Nazi 173
intention; intentionality 3, 4, 7, 22–5, 33, 35, 50–1, 60, 75–8, 80, 82–5, 87–90, 92, 97–9, 107, 114, 126, 159
intervention (in nature) 47, 72, 82, 85–6, 91–4, 107, 108, 117
invasion; invasive (of species) 139, 141, 143–6, 149
Israel; Israeli 21, 36, 140
Italy 9, 38–41, 66

J

Jaskot, Paul 60
Jenkins, Willis 102–03
Jew; Jewish; Jewry 2, 4, 8, 9, 14–15, 18, 21, 24, 35, 36, 38–42, 48–9, 55–7, 65–8, 124–5, 152, 155–6, 160, 173, 176, 180–1
Johst, Henns 18, 154
Jordan III, William 139–41, 145

K

Kammler, Hans 172
Kant, Immanuel 29, 30, 125
Keeling, Paul 100–04, 126
Kidner, David 126–6, 128, 131
killing *see also* murder 6, 15, 35, 41, 51, 56, 58–60, 62, 166–7, 169, 170
Kranzberg, Melvin 51
Krieger, Martin 18–19

L

labor *see also* slave 170, 172–3, 16, 25, 31, 62, 63, 65
  camp 15, 57, 60–2
landscape 1, 3, 10–12, 14, 16, 18, 26, 34, 43–5, 51, 75, 78, 93, 104, 107–08, 115–16, 120–4, 134, 139, 141, 146–7, 149–61, 164, 180, 181
Lange, Willy 152

language; linguistic 40, 53, 87, 95, 99, 101–05, 114, 123, 125–7, 133, 135, 138, 145, 149
Leiss, William 27
Lekan, Thomas 151
Lewis, C.S. 21
Levi, Primo 35
liberation *see also* freedom 7, 9, 15, 27–36, 42, 49, 68, 112, 115, 122, 125, 134, 178, 181
Lifton, Robert J. 165–6, 168, 174
Light, Andrew vii, 67, 79–81, 92
literature 10, 122, 130, 144
Lithuania 124_5
Lo, Yeuk-Sze vi, 83–8, 91–2
Lublin 8, 14–15, 59

M

machine; machinery 58, 62–4, 164
Majdanek 8, 10, 14–16, 22–4, 27, 34, 36, 40, 42, 49, 50, 59, 82, 181
malnutrition 15, 57, 169
Malpas, Jeff 44
management
  business 51, 62–4, 164–5, 168, 172–5
  of nature 2, 7, 9, 12, 17, 19, 47, 50, 65, 75, 93, 102, 106, 113, 122, 123, 132, 139, 149, 153, 157
Marcuse, Herbert 27–9, 32, 36
Marx, Karl; Marxism 53, 119
Maser, Chris 18–19
mastery *see also* domination; power 2, 10, 12, 78, 94
materialism 29, 128, 151–2
mathematics 43–4, 131–2
Mauthausen 60–1
McKibben, Bill 113
meaning 3, 4, 6, 7, 9–11, 17, 27–8, 30–1, 34–6, 40, 42–6, 50, 56, 66, 73, 77–8, 83–4, 86, 89, 91, 93, 95–6, 98–101, 104, 118, 121–3, 125, 127, 130, 134, 138, 141–4, 146, 150–1, 160, 166, 168–9, 181–2
mediation 30–1, 94, 114, 119, 125
memory 35, 68, 121, 181–2

Index

Merchant, Carolyn 27
Merleau-Ponty, Maurice 129
Mesthene, Emmanuel 51
metaphor 10–11, 22–3, 35, 102–03, 114, 120, 123, 125, 132–5, 139–41, 145, 152
metaphysics 28, 30–2, 34, 105
Meyer, Konrad 154
migration (of nonhuman species) 23, 141–8
Mill, J.S. 95
mining 21, 47, 48, 72–4
modernism 63
modification (of nature by humans) 18–19, 25–6, 28, 33, 36, 48–50, 65, 74–9, 83–4, 86–9, 97–9, 105, 112–17, 122, 125–6, 129, 132
morality *see also* ethics; values 4, 10, 12, 23, 28–30, 34, 36, 51–7, 60, 61, 63–5, 67, 75, 82, 91, 93, 99, 120, 126, 128, 146, 147, 149, 159, 165–75, 178
Moriarty, Paul 104, 142–3
Moses, Robert 54
mountain 47–8
Muir, John 122
Mumford, Lewis 51
murder *see also* genocide; killing 3, 6, 8, 9, 15, 49, 56, 58, 72, 74, 155, 166, 176
museum 16, 39, 41, 67, 132
Mussolini, Benito 41
myth 11, 121–5, 129, 130, 143

N

narrative 11, 43, 45–8, 65, 119, 121–4, 143, 144, 159, 170, 176–7
native
  culture; people 121–2, 139–40, 143, 155
  species 12, 139–49, 152–3, 156, 158, 160
nativism 12, 138–41, 145–9, 154, 156–7, 160

Nazi 2–4, 8–13, 14–18, 21, 27, 32, 34–5, 41–2, 52, 55–7, 60–1, 64–8, 72, 112, 138–42, 146, 149–61, 164–78, 180,
neutrality 160, 164, 169, 52
  of technology 10, 50–7, 61–2, 64–5, 167, 168–72, 174–8
New Jersey 2, 71
New York 2, 54, 71
New York City 1, 54, 76, 101, 126
nostalgia 1476
noumenal world 29–30

O

O'Brien, William 145–6, 148
objectivism; objectivity 43–4, 67, 127–8, 143
Operation Reinhard 15
opportunism 144, 150
oppression 5–7, 9, 11–13, 21–2, 24, 26–7, 30–1, 34–5, 50, 52, 60, 65, 112, 115, 120, 122–4, 138, 140, 164, 171, 178, 181–2
"otherness" 105, 129

P

Packard, Steve 90
Palestinian 21
panda 84
parent 31–1, 40, 90–1
park 1, 54, 85, 131, 133–4
  national 32, 122
peace 6, 8, 36, 112, 130, 138, 178, 181
perception 16, 29, 35, 44, 49, 71, 143, 166
Peretti, Jonah 142–3, 146
Peterson, Anna 116–18
Pinchot, Gifford 122
Pitt, Joseph 51
place, experience and importance of 16–17, 24, 35, 42–50, 66–8, 122, 130, 146–9, 181–2
planting 1, 2, 6, 18, 21, 25, 26, 34, 50, 71, 74, 80, 92, 98, 107, 113, 132–3, 146, 154, 181

Index

Plato 130
Plumwood, Val 103
policy
   environmental 9–12, 18-20, 32, 34, 45, 66, 72, 73, 78–79, 94, 97, 99–108, 114, 116, 127, 140, 143
   Nazi 11, 19, 32, 34, 42, 48, 55, 57, 60, 138, 140, 146, 149–57, 159–61, 172
politics 4, 10, 19, 31, 47, 52, 54–7, 60–4, 76, 94, 105, 118, 145, 50–3, 157, 159, 167–70, 172–8
Pollan, Michael 139–41, 146
pollution 20, 24, 78, 82, 93, 97–9
population *see also* community
   human 4, 15, 20, 41, 57, 59, 63, 71, 116–17, 139, 154, 155–7, 160, 161
   nonhuman 25, 85, 141, 143
poststructuralism 114, 116
power *see also* domination; mastery 2–4, 6, 8–12, 14, 16, 18, 19, 21–4, 26, 25, 36, 40, 42, 49, 54–5, 60, 78, 82, 94–5, 105–06, 114, 123, 124–5, 132, 135, 144, 152, 157–8, 160, 172, 181
pragmatism 30, 34, 61, 63, 73, 107, 114, 115
prejudice 21, 145
preservation *see also* conservation
   of culture and cultural sites 45–6, 67–8, 147
   of nature 3, 9, 10, 19, 29, 32, 34, 45–7, 58, 67–8, 73, 74, 79, 94–5, 100, 102–06, 108, 113, 116, 122, 124, 127, 129, 135, 141, 147–8, 150, 160
Pressac, Jean-Claude 58, 169, 176
Preston, Christopher 44
primitive 116, 134, 143
"productivism" 62–3
profit 62, 76, 152, 153, 173
progress 7, 82, 120, 123, 131
protection
   of culture 130, 140
   of nature 9, 10, 18, 25–6, 34, 71–3, 79, 94, 103–04, 106, 113, 116–17, 127, 149–52
Prüfer, Kurt 58
purity 3, 11, 12, 93, 106, 113–14, 127, 139–41, 146–9, 152–3, 157–8, 160–1, 174

R

racism 21, 116, 140–1, 145, 147–9, 152–3, 155, 161, 173
railway; railroad 54–5, 57, 59–60
rationality 7, 22–3, 61, 133, 140, 145, 147, 174
realism 99, 103–04, 125
recreation 1, 19, 32, 34, 153, 159
redemption 22, 131
Rehmann-Sutter, Christoph 44, 46
relationship
   between humans 7–8, 17, 30, 31, 36, 119
   human with nature 1, 3, 7–8, 11, 16–17, 22, 29, 31, 36, 44, 45, 53, 65, 79–80, 99, 117, 121–3, 126–7, 130, 154, 159, 181
religion 22, 31, 67, 121
remediation (of damage to nature) 99
resistance 4, 6, 7, 12, 13, 29, 73, 112, 120, 133, 134, 138, 161, 164, 178
resource, natural 1, 18–20, 24, 100, 133, 155
responsibility 3, 12, 92–3, 177
restoration, ecological 2, 9–12, 20, 25, 26, 33, 35, 42, 46–8, 50, 65–6, 71–108, 112–13, 116, 123, 126, 129, 138–41, 145–6, 149, 155–6, 158–61, 164
   focal 79–80
rhetoric 12, 138, 141, 145–6, 148, 150, 152, 153, 159, 160
rights 28, 155
RNG (Reich Nature Protection Law) 149–50
Rodman, John 28
Rolston, Holmes III vi, 45, 56

Index

Rome; Roman 41, 121, 130
Rousseau, Jean-Jacques 133–4
Rwanda 21

S

sacredness 1, 121
Sagoff, Mark 144–5, 147
Sale, Kirkpatrick 42
Sammons, Jack 170–2, 174–5, 178
sand *see also* beach 25–6, 32–4, 71–2, 83, 94, 107–08, 180–2
    dune 25, 34, 71–2, 74, 78, 83, 94, 107–08, 181
    mining 25, 72, 74
Sandy, Hurricane 2, 71–2, 106, 181
sapling 1, 3, 4, 6
savanna 21, 26, 90
Schama, Simon 10–11, 120–5, 129
science 3, 4, 9–12, 17, 18, 21, 25, 26, 33, 36, 44, 47, 50–1, 55, 65, 76–9, 81, 83, 92–3, 99, 104, 113, 115–20, 124–9, 131, 132, 140–5, 149, 152–3, 164–6
sea wall 34, 72, 107
Seifert, Alvin 151–4
Sequoia, Giant 106, 122
Sherwood Forest 121–2
Siipi, Helena 96–9, 106–07
Singer, Peter 28
slave; slavery 15–16, 21, 31, 57, 62–3, 65, 170, 173
Smith, Mick 118–19
snow fencing 34, 71, 83, 107
social constructivism *see* constructivism
Socrates 130
Soper, Kate 99–100, 103–05, 126
soul 131, 151
Speer, Albert 55–7, 64–5, 168–72, 174, 176, 178
spirit; spiritual 2, 19, 42, 62, 157, 171
SS 12, 15, 18, 58–63, 169–77
Stephens, Piers H.G. vi, 157
Stone, Christopher 28
storm 1, 2, 24, 71, 181

structure; structuring 19, 23, 30, 33, 34, 40, 44, 52–6, 61, 64–5, 72, 82, 90, 104, 114, 126, 127, 153, 158, 160, 175
subject; subjectivity 7, 23, 35, 43, 44, 67, 90, 127
    nature as 7, 22, 23–4, 27–30, 34, 77, 95
Sylvan, Richard 81–2, 92
synagogue 39, 41, 66–8, 182
    Spanish, Venice 9, 38–40, 42–3, 48–9, 66–8, 180, 181
system
    human 12, 52–7, 61–5, 112, 115, 120, 166, 166–7, 170, 175–6, 180
    natural *see also* ecosystem 21, 22, 25, 26, 29, 33, 36, 44, 45, 47, 48, 50, 73–8, 80–4, 87–90, 92, 94, 98–100, 102, 116–17, 119, 128–9, 143, 144, 158

T

Taylor, Bron vi, 43
Taylorism 62, 173
technician 64, 165, 168, 170–2, 174, 176
technocrat 61–2, 167, 173
technology 1–13, 18, 21, 25–6, 32, 38, 47, 48, 57–7, 62–5, 72, 81, 83–6, 89, 92–3, 112–13, 115, 123–5, 128–9, 164–78
    neutrality of *see* neutrality
    philosophy of 43, 46, 79, 164
teleology 7
Third Reich *see also* Nazi 3, 4, 7, 9, 11, 12, 50, 55–7, 61, 64–5, 112, 149, 151, 153, 164–5, 167, 172
Thompson, Janna 45
timber 1, 19, 25, 120, 123, 129, 132
Todt, Fritz 152
tolerance 1, 4, 139, 143
tool 53–4, 57, 63–5, 128, 167, 175
totalitarianism *see also* Nazi 56, 170
tourism 39, 68, 122, 132
tradition; traditional 3, 7, 10, 11, 51–2, 54, 56–7, 60, 64, 116, 121, 131, 153, 168–70, 174–6, 178

Index

transcendence 11, 30, 46, 89
Treblinka 14, 15
Tuan, Yi-Fu 44

U

Uekoetter, Frank 150, 152, 154–5, 159
urban 15, 54, 59, 113, 133–4, 168, 181

V

value 1, 7, 9, 10, 22, 23, 35, 42–8, 50–5, 66–7, 73–80, 86, 90–1, 95–6, 100–02, 104, 114, 118–19, 121, 125, 146, 156, 167–9, 175, 176
values *see also* ethics; morality 10, 13, 46, 52, 53, 55, 57, 60–5, 72, 105, 112, 116–19, 120, 122, 127, 143, 149, 151, 164–5, 168, 172–7
van Pelt, Robert Jan 17, 58, 154, 169, 176
vegetation 8, 14, 23, 25, 26, 34, 50, 71, 124
Venice 9, 38–43, 48–9, 66–8, 180
violence 8, 155
Vogel, Steven vi, 29, 83, 88–96, 100–02, 112–18
volksgemeinschaft 151–3, 158

W

war 21
  Second World 2, 6, 8, 14, 15, 21, 41, 48, 61, 64, 121, 167
Warsaw 2, 8, 14, 16, 22, 24, 26, 34, 36, 40, 42, 48–50, 82, 124, 181

weather 2, 23
welfare
  animal 141
  human 20
Wewelsburg 60
Wickop, Walter 155
Wiepking-Jürgensmann, Heinrich Friedrich 154–5
wilderness 1, 18, 32–4, 42, 74, 85, 100, 102, 104, 106, 113, 116–17, 121–3, 131, 139, 153
wildness 91–2, 102, 134
Winner, Langdon 52–5, 57, 64, 168, 175
Wisselinck, Kurt 172–3
Wittgenstein, Ludwig 53, 101, 126
wolf 85–6
wonder 1, 121, 181
wood *see* timber
Woods, Mark 142–3
wound 2, 9, 22, 24, 27, 49

X

xenophobia 141, 145–7

Y

Yellowstone 85, 144
Yugoslavia 21

Z

Zeller, Thomas 151–2
Zyklon-B 58–9

About the Author

Eric Katz (B.A. Philosophy, Yale; Ph.D., Boston University) is Professor of Philosophy and Chair of the Department of Humanities at the New Jersey Institute of Technology. He is the author of *Nature as Subject: Human Obligation and Natural Community* (1997), winner of the CHOICE book award for "Outstanding Academic Books for 1997." He is the editor of *Death by Design: Science, Technology, and Engineering in Nazi Germany* (2006). He has co-edited the collections *Environmental Pragmatism* (1996, with Andrew Light); and *Beneath the Surface: Critical Essays in the Philosophy of Deep Ecology* (2000, with Andrew Light and David Rothenberg); and the textbook *Controlling Technology* (2nd edition, 2003, with Andrew Light and William Thompson). He has written over fifty journal articles, book chapters and book reviews. He was the Book Review Editor of *Environmental Ethics* from 1996 to 2014, and was the founding Vice-President of the International Society for Environmental Ethics. From 1991 to 2007 he was the Director of the Science, Technology and Society (STS) programme at NJIT.

www.ingramcontent.com/pod-product-compliance
Lightning Source LLC
Chambersburg PA
CBHW021828300426
44114CB00009BA/363